GCSE
Biology

TEACH YOURSELF BOOKS

GCSE Biology

Sue Taylor

TEACH YOURSELF BOOKS
Hodder and Stoughton

First published 1987
Reissued 1989

Copyright © 1987
Sue Taylor
Illustrated by
Sue Taylor and Joanna Hunt

No part of this publication may be reproduced or transmitted in any form or by any means, electronically or mechanically, including photocopying, recording or any information storage or retrieval system, without either the prior permission in writing from the publisher or a licence, permitting restricted copying, issued by the Copyright Licensing Agency, 33–34 Alfred Place, London WC1E 7DP.

British Library Cataloguing in Publication Data

Taylor, Sue
GCSE biology.—
(Teach yourself books).
1. Biology
I. Title
574   QH308.7

ISBN 0 340 51282 2

Printed in Great Britain for
Hodder and Stoughton Educational,
a division of Hodder and Stoughton Ltd,
Mill Road, Dunton Green, Sevenoaks, Kent,
by Richard Clay Ltd, Bungay, Suffolk
Photoset by Rowland Phototypesetting Ltd,
Bury St Edmunds, Suffolk

CONTENTS

Introduction vi

1 The Diversity of Organisms 1
1.1 Organisms and classification 1.2 The plant kingdom 1.3 The animal kingdom *Sample questions*

2 Organisms and the Environment 15
2.1 The biosphere 2.2 Relationships between organisms 2.3 Diseases and defence 2.4 Cycles of matter 2.5 Soil 2.6 Practical ecological study 2.7 Population size and control 2.8 Man and the environment *Sample questions*

3 The Individual Organism 43
3.1 Studying cells 3.2 Specialisation 3.3 Cells in action 3.4 Living things need food 3.5 Plants and food: photosynthesis 3.6 Animals and food 3.7 Transport in flowering plants 3.8 Transport in mammals 3.9 Respiration 3.10 Gas exchange 3.11 Excretion 3.12 Sensitivity 3.13 Response and Structure 3.14 Co-ordination 3.15 Homeostasis *Sample questions*

4 Development and Reproduction 131
4.1 Chromosomes and cell division 4.2 Growth and life cycles 4.3 Asexual reproduction 4.4 Sexual reproduction 4.5 Inheritance 4.6 Variation 4.7 Selection and evolution *Sample questions*

Answers 174

INTRODUCTION

You have chosen to study the fascinating subject of biology and should have had the chance to observe living things and carry out experiments to discover more about them. This book is to use as you prepare for your GCSE examinations.

It starts by explaining what is involved in the examining system and what to expect in your examination papers. It gives advice on answering different types of question and on your approach to revision. The next few pages should make things clearer and give you confidence to tackle successful revision. It is worth spending time reading them. I hope that by the time you reach the last page, you really are ready for examinations. Good luck!

What is GCSE?

The General Certificate of Secondary Education replaces the GCE 'O' level, CSE and Joint 16 plus examinations. If you entered secondary school any time after September 1983, you are likely to take GCSE, whether you are studying in a maintained school, an independent school, a College of Further Education or privately. The standards expected are at least as high as in the previous examinations, but GCSE is a single system of education with a single scale of grades.

GCSE will test not only your memory and the orderly presentation of facts, but other skills and the ability to apply knowledge. Some assessment will therefore have taken place in school or college during ordinary lessons and will be considered with the results of the formal examinations at the end of the two year course. All courses and examinations follow nationally agreed guidelines, known as 'national criteria'.

Grading
You will each be measured against defined yardsticks called 'grade criteria' which spell out what you need to know, understand and be able to do in order to achieve a particular grade. Successful candidates will be awarded a grade from A to G. (This helps employers and institutions that need to compare you with other candidates from different areas and with different backgrounds.) Unsuccessful candidates will be ungraded and will not receive GCSE certificates. (It is possible to resit GCSE examinations and to attempt different subjects at more than one sitting.)

Core content
Each of the Examining Groups publishes its own syllabus for each subject, including biology. However, at least sixty per cent of the subject matter contained in each has been agreed nationally. This is called the 'core content' and is studied by everyone attempting GCSE examinations. Topics are classified under four main 'themes' or headings:

Introduction vii

1 Diversity of organisms
2 Relationships between Organisms and with the Environment
3 Organisation and Maintenance of the Individual
4 Development of Organisms and the Continuity of Life.

The four main chapters of this book correspond to these four themes.

The examining groups
There are six examining groups incorporating the old examining boards. Before you start revision it is important that you find out which group you are being examined by. *This will not necessarily be the same for all your subjects.*

The examinations
The assessment pattern differs between examining groups. Check details for your group so that you know the number and type of examinations to expect. This information is given at the beginning of each printed syllabus, which you can obtain from school, college or your local library. Note that a certain percentage of your final mark is based on work that you have done already. It will include an assessment of your practical skills and you may have had to submit some of your course work. With your teacher you will have decided which grade you are aiming for. Remember that this may affect the particular examination papers you will be sitting. e.g.

Paper 1:	15–20 short-answer questions + two structured questions (all compulsory)	(80%)
Paper 2:	Centre (school/college) assessment of certain practical skills	(20%)
Paper 3:	structured questions (all compulsory)	

Papers 1 and 2 are compulsory and enable you to achieve any grade up to and including C. Paper 3 is optional and enables you to raise your grade by one or two. Thus, grades A and B can be achieved only by candidates taking all three papers.

Types of examination question

Read the sections which apply to your examining group's papers.

Multiple choice questions
These are sometimes called objective questions or fixed response questions. You are asked to choose the best answer from a choice of four or five which are given, e.g.

viii *Introduction*

1 Which of the following plants does not contain chlorophyll?
   A mushroom   **B** dandelion
   C hair moss   **D** pine tree

The correct answer to this question is A and you would be expected to indicate this by filling in the correct box or brackets, using a pencil, or on a separate answer sheet. This is marked by computer, e.g.

   1 A(−)   B( )   C( )   D( )
   2 A( )   B( )   C( )   D( )
   3 etc.

There is only one correct response and you would be marked wrong if you filled in more. If you do not know the right answer, you can start by eliminating any you know are wrong and choose one of the remaining answers, with a better chance of gaining a mark. Never leave a blank! Most mistakes are made by misreading the often lengthy questions, so take great care and double check each answer if time allows. Do not spend a long time on any one question you find hard, unless you have tried the rest!

Structured questions
These are sometimes called short-answer questions. You are asked to give answers about topics you have studied in spaces provided on the question paper. The size of space and the number of marks allocated (usually shown to the right of the question in brackets or a margin) will indicate whether you are expected to answer in a word, a sentence or a paragraph, e.g.

   (*a*) Name a hormone produced in the body of a mammal   (1)
        Answer: Adrenalin
   (*b*) Name the endocrine organ(s) that make(s) this hormone   (1)
        Answer: Adrenal glands
   (*c*) Describe two effects of this hormone in the body   (2)
        Answer: It increases heart beat and raises blood glucose level

Try to provide one fact for each mark allowed and do not cram essays between the dotted lines! Always read all of the sections before starting to answer. This ensures that you do not waste time repeating yourself and that you are gaining maximum marks. For example, you may have picked a different hormone to answer part (*a*) above and find yourself unable to think of a second effect it has for part (*c*). Diagrams, tables, graphs and photographs are often used in these questions. Do not ignore any piece of information given, as it will help you to arrive at the right answer.

Analysis of data questions

These questions test your understanding and your ability to use information. They may be about topics that you have not studied, or about experiments with unexpected results. Diagrams, recordings, tables of data and graphs may be used, and you will be expected to discuss or interpret them. In addition you may have to add facts from your syllabus, make calculations (such as finding averages) and suggest further experiments to resolve unknowns, e.g.

The following table includes information about percentages of some of the substances present in blood plasma, the filtrate into kidney tubules (nephrons) and urine.

Substances	% in plasma	% filtrate in nephron	% in urine
Protein	7.0	0	0
Glucose	0.1	0.1	0
Creatinine	0.001	0.001	0.075
Urea	0.03	0.03	2.0

Answer the following questions.

(a) Which substance is not filtered from the blood plasma? (1)
Answer: protein
(b) Name an example of this substance found in the plasma. (1)
Answer: fibrinogen
(c) Give one reason why filtration does not occur with this substance. (1)
Answer: molecules are too large
(d) Which substance filtered into the nephrons is taken back into the blood? (1)
Answer: glucose
(e) State your reasons based on the evidence given. (1)
Answer: there is the same percentage in plasma and nephrons but none in the urine

Always show any calculations you make; marks are given for these, as well as the final answer. Use your common sense and again be guided by the space allowed for answers and the mark allocation.

Essay-type questions

These are sometimes called free-response questions. They are carefully worded to direct your answer, so pay attention to every word in the question, and do not twist questions to suit your knowledge! For example,

> Describe the process of digestion between the oesophagus and
> colon in mammals. (25)

You should notice the following points straight away:
1. digestion i.e. *physical and chemical breakdown of food, not absorption.*
2. between the oesophagus and colon i.e. *not in the mouth or beyond the colon.*
3. in mammals i.e. *not just humans but including cellulose digestion in ruminants.*

Always plan your answers. You will be less likely to leave out key facts and more likely to arrange information logically. In general there will be one mark for each relevant fact and none for waffle! Simple, clear, labelled diagrams are of use as long as they do not repeat written information. Essay-type questions can be of traditional style, as above, where you will have to decide on your own framework, or more structured and made up of several parts, where a framework is given to you, e.g.

> Explain the role of muscle action in
>
> (*a*) accommodation in the mammalian eye (5)
> (*b*) controlling the amount of light entering the eye (5)
> (*c*) a kneejerk response (5)
> (*d*) peristalsis in the gut (5)
> (*e*) childbirth (5)

In this type it is easier to see where marks are given.

Revision for success!

Having decided on realistic goals, how are you going to achieve them? If you have understood how GCSE is assessed you should realise that continual effort is needed, particularly in practical sessions. Your teachers will have sent the results of assignments that you have completed in lessons (and possibly some of your course work) off to the Examining Groups. The rest depends on the examinations themselves. You have got to do them: you may as well do them well!

You must want to succeed. By reading this you are showing willing and now you must be prepared to work. You must use common sense and honesty with yourself as you decide on the best revision plan for you, but first, consider all the things that you have done already.

Introduction xi

Your teacher has
done his or her best by . . .

You should have
done your bit by . . .

          listening, thinking, questioning and doing homework

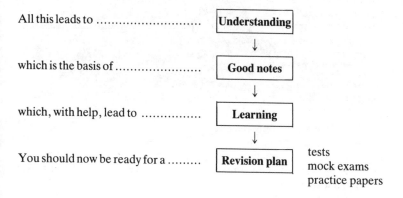

If you have not reached this stage, don't panic! Revision means looking at again, so it is usually easier to start with your familiar notes in your own handwriting and if you have understood first time round, it should be easy! First of all, tidy your school or college notebooks or files and check that your notes are complete. If not, this book will be a substitute for any gaps or weak places.

Next, you need to know which topics you must learn for your Examining Group. With a copy of your syllabus in front of you, record clearly the topics to revise. To help you in this, this book is arranged, like your syllabus, under the four major 'themes' each divided into a number of topics. All you have to do is mark off the relevant topics; other topics can be ignored.

xii *Introduction*

An understanding of biology requires an overall picture of living things and for this reason you are advised not to leave chunks of the syllabus out of your revision. Moreover, multiple-choice and short-answer questions make up a large part of examination papers and demand precise knowledge of all parts of the course. (There are GCSE-style questions at the end of each theme for you to consult.)

Plan how many topics you need to revise each week to finish the syllabus before the examination and draw up your revision plan. Here are some final points to consider as you plan:

DON'T

1 try to work for too long too late at night;
2 play loud music or watch TV;
3 get up to make countless drinks and snacks;
4 revise by simply reading and rereading notes or by copying out notes again and again.

Introduction xiii

DO

1 find out how long you can hold your concentration;
2 find somewhere quiet to work;
3 promise yourself a treat when you've finished an assignment;
4 revise by sensible methods such as:

(a) Underlining headings and key words
(b) Copying out key facts
(c) Drawing and labelling simple diagrams
(d) Learning parrot-fashion, i.e. by repeating something over and over again. This works well but doesn't help understanding. You can't learn a wide biology course this way but it may be useful for some difficult definitions, labels and equations.
(e) Using patterns or flow charts, which can be of great help because they are easier to visualise than blocks of words. You could make a summary of each topic like this.
(f) Testing yourself without roping in friends or family! This book contains many key diagrams with labels beneath them so you can test yourself by holding a piece of paper over the 'answers'. Tables can be treated similarly by covering a column or row.
(g) Using mnemonics, which are words, sentences or rhymes that act as memory triggers. They can be personal (including friends' initials etc.), funny or very rude, as long as they work and you don't write them on your answer sheet! Some examples are used in this book, e.g. the three conditions needed for most seed germination are **MOST**! – **M**oisture, **O**xygen and **S**uitable **T**emperature.
(h) Practising questions from past papers, which will
   (i) check that you know the facts,
   (ii) help you to become familiar and confident with different types of question,
   (iii) help you to get your timing right.

THE DIVERSITY OF ORGANISMS
1.1 Organisms and classification

Biology is the study of organisms or living things. What distinguishes these living things from non-living things? Each *organism* demonstrates *all seven* of the following characteristics:

1. Obtains food (nutrition),
2. Removes waste (excretion),
3. Grows,
4. Adds more to its own kind (reproduction),
5. Needs energy (from respiration),
   Is
6. Sensitive, and
7. Moves.

Non-living things may show *some* of these characteristics (e.g. a train moves and crystals grow), but only organisms fulfil all seven.

The need for classification

There are over two millon known species of plant and animal in the world and more are still being discovered. No one can hope to know all of these, so they are grouped. The branch of biology concerned with grouping and naming organisms is called *taxonomy* or *systematics*.

The most commonly used system was devised by Linnaeus in 1735. It is called the *Binomial System* because it gives a two-part biological name to each species. Once again there are *seven* things to remember when memorising the levels of classification:

Level of classification	Examples	
Kingdom	Animal	Plant
Phylum (plural Phyla)	Chordata	Spermatophyta
Class	Mammals	Gymnosperms
Order	Primate	Coniferales
Family	Hominidae	Pinaceae
Genus (plural Genera)	*Homo*	*Pinus*
Species (plural Species)	*Homo sapiens*	*Pinus sylvestris*

There is usually only one kind of organism in a species, but a few examples such as dogs and roses have several *varieties*. Notice that the genus name is written with a capital letter and the species name is written without a capital letter. Both parts are in Latin so that they can be used all over the world. They are printed in italics or underlined if written (which is important in examinations). Each country also gives each species it recognises a *common name*. The examples above are *Man* and *Scot's Pine*.

Organisms and classification

Keys

In order to find the way into the right place in a classification system, a *key* is often used. This enables us to identify and name an organism we are not familiar with. Keys are constructed by looking at the differences in *external features* of a collection of organisms, finding one that splits the group into two, and repeating the process to give smaller and smaller groups. When only one remains in each group, it can be given its name. Here is a simple example. Follow it through carefully.

Tree leaf example

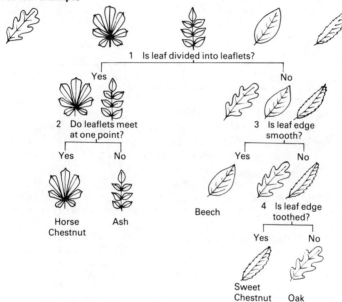

Usually, keys are then reorganised into a series of numbered questions and the user answers 'yes' or 'no' or selects one of two contrasting descriptions that suits the unfamiliar organism:

1	Is leaf divided into leaflets?	If yes, go to 2.
		If no, go to 3.
2	Do leaflets meet at one point?	If yes, it is *Ash*.
		If no, it is *Horse Chestnut*.
3	Is leaf edge smooth?	If yes, it is *Beech*.
		If no, go to 4.
4	Is leaf edge toothed?	If yes, it is *Sweet Chestnut*.
		If no, it is *Oak*.

THE DIVERSITY OF ORGANISMS
1.2 The plant kingdom

A key to the plant kingdom could be set out as follows:

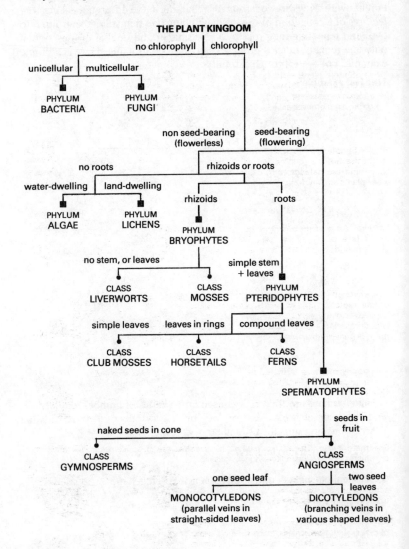

Plants

Seven phyla are usually recognised:

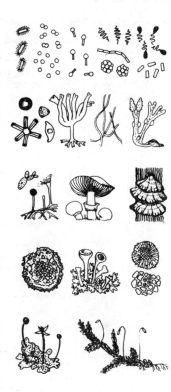

Bacteria Smallest and simplest organisms (about 0.001 mm). Have no chlorophyll but some make their own food. Occur in water, air, soil and other organisms.

Algae Simple plants with no roots, stems or leaves. Live in water. All have green chlorophyll for *photosynthesis* (see pp. 67–72). Some have red or brown pigments as well.

Fungi Simple organisms which do not photosynthesise but are usually *Saprophytic* (see p. 59). Some are parasitic and cause disease in other plants and animals.

Lichens An alga and fungus living in *symbiosis* (see p. 20). (Very sensitive to atmospheric pollution and often used as an indicator.)

Bryophytes (liverworts and mosses) Have no proper roots but rhizoids. Have simple stems and leaves or thallus (plant body). Found mainly in damp places. Have *sporophyte* and *gametophyte* generations, alternately.

Pteridophytes (club mosses, horsetails and ferns) Have proper stems, roots and leaves. Produce reproductive spores in autumn. Found mainly in damp places. Exhibit alternation of generations.

Spermatophytes Have proper stems, roots and leaves. Produce seeds. This phylum of organisms includes the following two classes:

Gymnosperms (conifers) Large plants with cones bearing seeds for reproduction. Needle-like leaves that are retained in winter.

Angiosperms (flowering plants) Wide range of plants with flowers and fruits bearing seeds for reproduction. Range from small herbs to massive trees.

5 The plant kingdom

Plant examples

Fungi

Mucor (Bread mould)   **Saccharomyces (Yeast)**

Algae

Spirogyra   **Chlamydomonas**

- **A** mycelium
- **B** aerial hyphae
- **C** sporangium
- **D** spores
- **E** nucleus
- **F** membrane
- **G** cell wall
- **H** cytoplasm
- **I** chloroplast
- **J** vacuole
- **K** nucleus
- **L** flagellum

Flowering plants (angiosperms)

Whole plant   **Half flower**

Terminal bud develops into flower during the reproductive phase.

- **A** node
- **B** internode
- **C** leaf
- **D** stem
- **E** tap root
- **F** lateral root
- **G** root hair
- **H** petal
- **I** sepal
- **J** receptacle
- **K** pedicel
- **L** anther
- **M** filament
- **N** stamen
- **O** stigma
- **P** style
- **Q** ovary
- **R** carpel

The plant kingdom 6

Useful plants

Bacteria and fungi: Saprophytic species are responsible for leaf and corpse decay. Others are used in production of cheese, vinegar, antibiotics and yoghurt.
Yeasts (fungi): Used in baking and brewing industries.
Lichens: Indicators of atmospheric pollution.
Algae: Produce *alginates* for use in the food industry. They suspend solids in ice cream and drinking chocolate and keep the head on beer.
Flowering plants: As they *photosynthesise* they trap sunlight energy and store it as chemical energy in *food*, usually sugars or starches. These form the *staple diet* of many animals, including man (see below).

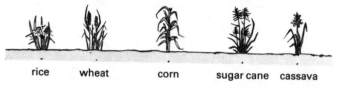

rice   wheat   corn   sugar cane   cassava

Many others are eaten as 'fruit' and 'vegetables'. Oxygen is produced at the same time and helps to balance the gases in the atmosphere. Flowering plants have many other products.

(a) *Fibres:* e.g. cotton from cotton plant flowers, linen from flax stems, sisal and jute from leaves;
(b) *Rubber* is made from latex, a milky juice in stems of several species;
(c) *Gums and resins* (e.g. from pines) are used for adhesives (glue), varnishes, paints, sealing wax and sizing (stiffener) in paper;
(d) *Drugs:* e.g. aspirin, originally from willow bark, rotenone from *Derris*.
(e) *Other chemicals:* e.g. insecticide from *Pyrethrum* flowers.
(f) *Wood and its products:*

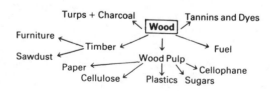

THE DIVERSITY OF ORGANISMS
1.3 The animal kingdom

A key to the Animal Kingdom could be set out as follows:

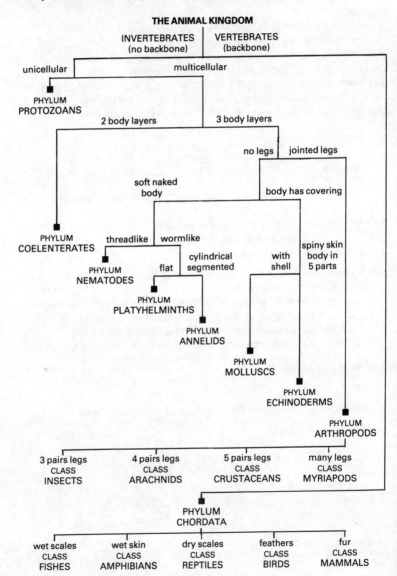

Animals A: Invertebrates

Eight phyla are usually recognised:

Protozoans Microscopic animals, one cell only. Aquatic or parasitic, some causing diseases.

Coelenterates Have a hollow body with two layers of cells. The mouth is the only opening and is surrounded by tentacles bearing sting cells. Most live in the sea.

Nematodes (roundworms) Body elongated and thread-like (round in cross-section). Some are parasitic causing diseases.

Platyhelminths (flatworms) Body elongated and flat. Some live in fresh water. Most are parasitic, causing diseases.

Annelids Have cylindrical bodies made up of rings or segments, with *chaetae* (bristles).

Molluscs Unsegmented, soft-bodied with one or two shells, sometimes greatly reduced and inside the body.

Echinoderms Have a tough spiny skin and sucker feet. Body is divided into five parts. All live in the sea.

Arthropods This is the largest group of invertebrates. Have segmented bodies, exoskeletons of *chitin* and jointed legs. This phylum of organisms includes the following four classes:

Myriapods (centipedes and millipedes) Body with very many segments, each bearing one pair of legs.

Crustaceans Mostly aquatic. Some are very large. Five pairs of legs.

Arachnids Body in two parts, (cephalothorax and abdomen). Eight legs. No wings.

Insects Body in three parts (head, thorax and abdomen). Six legs, usually two pairs of wings in adult.

9 The animal kingdom

Animals B: Vertebrates

Chordata This phylum of organisms includes the following five classes:

Fishes Live in water. Have gills for gas exchange, slimy, bony scales on their skin and fins for movement. Cold-blooded (*ectothermic*).

Amphibians Most live on land but lay eggs in water. Have gills and lungs. Have moist skin without scales. Have fish-like young called tadpoles. Ectothermic.

Reptiles Most live on land where they lay eggs with leathery shells. Have lungs. Have dry skin with waterproof scales. Ectothermic.

Birds Live on land and in air. Lay hard-shelled eggs. Have lungs. Skin covered with feathers. Fore-limbs are wings for flying. Beaks adapted for feeding. Warm-blooded (*endothermic*).

Mammals Most live on land. Have lungs and diaphragms. Skin covered with fur. Young are fed on milk made in mammary glands. Endothermic.

Types of mammal

Group	Examples	Where found	Features
monotremes	duck-billed platypus	Australia & S. America	Lay eggs in burrows. Webbed feet for swimming and 'beak' for feeding
marsupials	koala bear, kangaroo	Australia	Give birth to small immature young which are kept in a pouch
placentals	human, cat, whale	throughout world	Develop fully inside female's uterus, getting food and oxygen through a placenta

The animal kingdom 10

Animal examples

Annelids
Earthworm

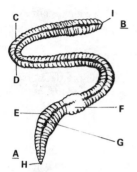

Insects
Honey Bee

A	anterior	F	saddle
B	posterior	G	segment
C	dorsal side	H	mouth
D	ventral side	I	anus
E	moist skin		

J	head	Q	wing
K	thorax	P	pollen basket
L	hairy abdomen	Q	jointed leg
M	compound eye	R	sting
N	cuticle		

Fishes
Mackerel

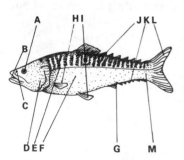

Birds
Sparrow

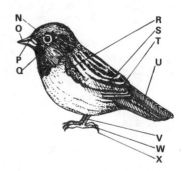

A	eye	H	pectoral fin
B	mouth	I	pelvic fin
C	nostril	J	dorsal fin
D	gill flap	K	ventral fin
E	lateral line	L	tail fin
F	scaly skin	M	tail muscle
G	anus/cloaca		

N	eye	T	wing
O	beak	U	tail (feathers)
P	nostril	V	leg
Q	ear	W	toe
R	covert feathers	X	claw
S	flight feathers		

11 The animal kingdom

Useful animals

Invertebrates
(a) *Pollination:* e.g. bee transfers pollen between crop plants;
(b) *Soil fertility is increased:* e.g. by earthworm;
(c) *Products:* e.g. silk moth caterpillar makes silk and bee makes honey;
(d) *Biological control:* e.g. ladybird eats aphids (greenfly) and praying mantis eats other pests;
(e) *Food:* e.g. squid, crab, lobster, prawns and shrimps.

Vertebrates Man has domesticated many species including cattle, horses, sheep, goats, chickens, geese, ducks and trout. These are used for:
(a) *Transport and ploughing:* e.g. horses;
(b) *Food:* meat is animal muscle;
(c) *Products:* e.g. milk, butter and cheese from cows and goats, eggs from chickens and ducks, manure from horses, glue from bones, leather from cows, wool from sheep;
(d) *Pets:* e.g. budgerigars, cats, dogs, gerbils, mice, horses;
(e) *Guarding property:* e.g. dogs;
(f) *Biological control:* e.g. cats or snakes controlling mice in barns.

Wild animals are also exploited: *Fishes* (e.g. herring, cod, plaice) provide food; *Whales* provide blubber for foods and cosmetics; *Seals* provide pelts; *Crocodiles* provide skins for handbags, shoes, etc; *Snakes:* some species provide drugs.

The importance of insects

The number of insects on Earth is enormous. The pie chart shows how many insect species there are in relation to other animal species.

Useful effects (see above)
(a) Pollination
(b) Products
(c) Biological control

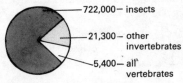

722,000 — insects
21,300 — other invertebrates
5,400 — all vertebrates

Harmful effects
(a) *Spread of disease:* e.g. mosquito spreads malaria, tsetse fly spreads sleeping sickness;
(b) *Crop spoilage:* e.g. locusts eat any crop, Colorado beetles eat potatoes;
(c) *Food spoilage:* e.g. housefly and bluebottle lay eggs in food, and maggots emerge;
(d) *Wood spoilage:* e.g. death watch beetle and woodworm attack buildings and furniture;
(e) *Clothes spoilage:* e.g. clothes moth eats wool;
(f) *Irritation:* e.g. blood-sucking flea and headlouse cause itching of skin.

1 THE DIVERSITY OF ORGANISMS
Sample questions

1. The system of classification used today was devised by
   A Malpighi   B Linnaeus   C Harvey   D Malthus (1)

2. Which of the following series places classifications in ascending order?
   1 species, 2 phylum, 3 genus, 4 class, 5 family, 6 order.
   A 1 3 5 6 4 2   B 4 2 5 6 1 3   C 5 6 4 3 2 1   D 1 2 6 5 4 3 (1)

3. Which of the following do not contain chlorophyll?
   A Pines   B Seaweeds   C Mosses   D Fungi (1)

4. Which of the following produce seeds with fruits?
   A Liverworts   B Conifers   C Angiosperms   D Ferns (1)

5. Which of the following are vertebrates?
   A Starfish   B Jellyfish
   C Dog fish   D Sea anemones (1)

6. Which of the following do *all* invertebrates have in common?
   A jointed legs   B exoskeletons   C no backbone   D antennae (1)

7. Which of the following is true for *both* bacteria and protozoa?
   A have cell walls   B all cause disease   C have nuclei
   D unicellular (1)

8. The number of legs possessed by arachnids (spiders) is
   A eight   B six   C four   D ten (1)

9. The most common invertebrate class is the
   A Myriapods   B Insects   C Crustaceans   D Arachnids (1)

10. Which of the following are endothermic?
    A Fishes   B Mammals   C Reptiles   D Amphibians (1)

11. Which invertebrate phylum contains individuals with hollow bodies with two layers of cells and tentacle-fringed mouths?
    A Nematodes   B Annelids   C Molluscs   D Coelenterates (1)

12. The front of an organism is known as its
    A dorsal side   B ventral side   C anterior   D posterior (1)

13. Which animal in the following list has compound eyes?
    A Honey Bee   B Man   C Owl   D Earthworm (1)

14. Another name for a flower stalk is
    A receptacle   B pedicel   C petiole   D lamina (1)

15. Which of the following is found in the root but not the stem of a flowering plant?
    A epidermis   B hair cell   C xylem   D phloem (1)

13 Sample questions

16 (a) Use the following key to classify the animals labelled A, B, C, D and E below. (5)

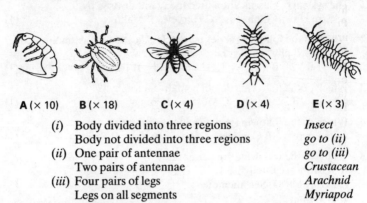

(i)	Body divided into three regions	*Insect*
	Body not divided into three regions	*go to (ii)*
(ii)	One pair of antennae	*go to (iii)*
	Two pairs of antennae	*Crustacean*
(iii)	Four pairs of legs	*Arachnid*
	Legs on all segments	*Myriapod*

Write your answers in the table below:

Animal	Group to which it belongs
A	
B	
C	
D	
E	

(b) The five organisms are all arthropods. List *two* features shown by all arthropods. (2)
(c) Organism A has × 10 written next to it. What does this mean? (1)
(d) List two *other* external features of organism C which could be used in a key. (2)

17 *Read the following passage and then answer the questions that follow:*
The Plant and Animal Kingdoms were defined before the microscope was in common use. There are already nearly 10,000 recorded species that are neither plant or animal or have features of both. Many biologists recognise four more kingdoms to accommodate examples like these.

Monerans These have rigid cell walls containing nitrogen, no proper nuclei, and no endoplasmic reticulum. Some are unicellular; others live in colonies.

Protists These live in water or damp places. They have nuclei. Most are unicellular; others live in colonies.

Fungi These include *True Fungi* and *Slime Moulds*. They have cell walls of *chitin* (not cellulose) nuclei and other membrane-bound organelles. Some are unicellular; many lack cross-walls i.e. are aseptate. They have no green chlorophyll and do not photosynthesise.

Lichens These comprise an *Alga* and a *Fungus* living in *symbiosis* (*mutualism*). They are land dwelling and have complex forms.

(a) Define the following words:
  (i) unicellular
  (ii) aseptate
  (iii) symbiosis (mutualism) (3)
(b) From your knowledge, which of the four kingdoms include
  (i) Protozoans and Simple Algae
  (ii) Bacteria and Blue-green Algae (2)
(c) Give *three* reasons why Fungi do not fit well into the Plant Kingdom (3)
(d) *Euglena*, a Protist, is illustrated below. Which of the seven labelled structures are usually considered plant features and which animal features? (7)

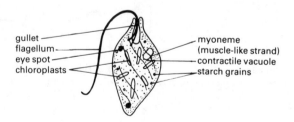

18 How has man found plants and animals useful to him? Give *named* examples. (20)

ORGANISMS AND THE ENVIRONMENT
2.1 The biosphere

Ecology is the scientific study of the relationships between organisms and their environment. The environment is made up of *biotic* (living) and *abiotic* (non-living) components.

The *biosphere* is the layer of the Earth's surface that supports the biotic component. It is separate from space but is greatly influenced from outside by solar radiation.

Height	Levels		Life Forms	Conditions
	stratosphere			
Altitude (km)	atmosphere	(km)		
16		9.5	bacteria, pollen	Low oxygen level high ozone level
12		8.2	birds	UV rays
		6.7	spiders, insects	
8	snow line			
	tree line	<6.1	most land plants + animals	
4				
0	sea level			maximum soil depth
depth (km)	hydrosphere (oceans)		seaweeds	light water
−4	earth's crust		inverte- brates	dark water
−8			fish	
−12				

Most organisms are found in a shallow layer of the biosphere. Biotic and abiotic components combine *in a particular area* to form an *ecosystem*. This is a self-sustaining unit of the biosphere. An example is an oak woodland. An ecosystem contains a number of *habitats*. A habitat is a particular type of place in which you find *communities* of organisms adapted to life there. An example is a rotting log, supporting a community of fungi, mosses and woodlice. *Niche* is the part that an organism plays in a habitat. For the example above, the woodlouse's niche is to break down dead timber and help to release minerals.

Stability of ecosystems

Simple ecosystems, i.e. those that involve few organisms, tend to be unstable. Changes in the environment lead to large changes in the numbers of the species present.

Complex ecosystems are more stable. The interaction between a larger number of species allows the system to buffer the effects of environmental change.

Causes of instability

Succession Organisms that are well-adapted become dominant for a while, and the process repeats itself as environmental change continues. Each recognisable stage is called a *sere* and the system reaches a *climax* when it ultimately stabilises; for example, is said that if bare soil was left untouched in England, beech and oak woodland would eventually take over (see below).

Plant succession

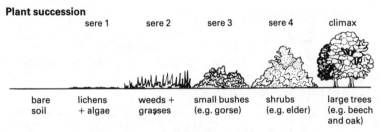

sere 1	sere 2	sere 3	sere 4	climax
bare soil	lichens + algae	weeds + grasses	small bushes (e.g. gorse)	shrubs (e.g. elder)	large trees (e.g. beech and oak)

Each sere in this plant succession would support different animal species.

Disease/catastrophes These can have sudden and devastating effects. *Myxamotosis* killed countless rabbits, grasslands were no longer grazed and shrubs invaded, completely altering the landscape. *Fire* can also have disastrous effects.

Pollution This can affect all parts of an ecosystem, but some species are more sensitive than others and this creates an imbalance. For example, the Industrial Revolution brought about the release of soot and chemical-laden smoke. This killed lichens and 'selected' dark-coloured prey animals that were better camouflaged (see p. 168). Pollution caused by oil spillage and acid rain has highly damaging effects.

ORGANISMS AND THE ENVIRONMENT
2.2 Relationships between organisms

Food chains

Green plants are called *producers* since they produce their own food. Animals are called *consumers* since they have to consume 'ready-made' food in the form of plants or other animals. A plant being eaten in this way forms the first *link* in a *food chain*.

A Link

grass ⟶ rabbit

An arrow indicates the direction in which food is passing.

A Food Chain (This example has two links in it.)

grass ⟶ rabbit ⟶ fox

Hence a food chain can be written:

producer ⟶ *primary* ⟶ *secondary* ⟶ *tertiary* ⟶ (and so on.)
            (1st order)   (2nd order)   (3rd order)
            consumer   *consumer*   *consumer*

Consumers are also grouped under:
(a) *Herbivores:* consumers that eat plants (e.g. rabbit)
(b) *Carnivores:* consumers that eat animals (e.g. fox).
(c) *Omnivores:* consumers that eat both plants and animals (e.g. human).

Remember that food chains, when written, should:

(a) always start with a producer;
(b) contain at least two links, but seldom more than four;
(c) always include arrows showing the direction that food is passing.

It is more difficult to think of longer food chains, so learn two – for example:

In a garden:
rose ⟶ greenfly (aphid) ⟶ ladybird ⟶ hedgehog ⟶ fox

In a freshwater pond:
algae ⟶ water flea ⟶ water beetle ⟶ frog ⟶ grass snake

Food pyramids

Pyramids of numbers As a general rule a primary consumer needs many smaller plants to feed on. A secondary consumer is usually a larger animal than its food, and hence needs to eat more than one if it is to stay alive. This relationship is drawn as a *pyramid of numbers*. For any particular food chain (e.g. wheat ⟶ mouse ⟶ hawk), the number of organisms at each position (sometimes called *trophic level*) in the food chain can be calculated. Then, a pyramid is constructed of blocks representing in size the numbers involved.

Pyramids of numbers

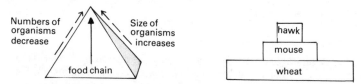

However, numbers in some food chains do not give pyramids:

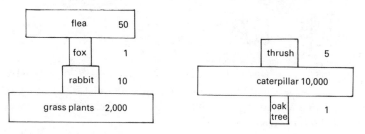

Pyramids of biomass These overcome the numbers problems. If the wet or dry mass of each trophic level is measured, the *biomass* (mass of living material) always gives a pyramid:

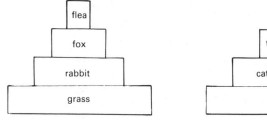

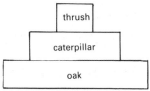

This is because the fleas each have very small mass.

This is because the oak tree has a very large mass.

19 Relationships between organisms

Pyramids of energy Energy enters an ecosystem when *sunlight energy* is trapped by green plants during *photosynthesis*. Plants use the chemicals which are produced (e.g. *sugar* and *starch*) as energy stores for themselves. The energy is only passed on, in food chains and webs, when the plants are eaten. Energy is lost at each trophic level and a pyramid is formed (see example below).

- 4% for growth
- 33% { converted to movement / used to maintain body temperature / released by respiration and digestion }
- 63% { lost as some grass is indigestible }

In general, only 1–10% of energy reaches the next trophic level. Thus, it is much more efficient for humans to eat plant crops. If plants are fed to animals that are to be eaten, a lot of energy is lost because of the extra trophic level.

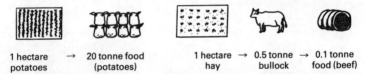

1 hectare potatoes → 20 tonne food (potatoes)

1 hectare hay → 0.5 tonne bullock → 0.1 tonne food (beef)

Such knowledge is important in famine relief work. It ialso goes some way to explaining why animal foods (e.g. meat, fish and cheese) are so expensive to buy. Energy costs money!

Food webs

In an ecosystem, there is rarely a single food chain. Instead there is a network of cross-linked chains. A producer such as grass may be eaten not only by rabbits, but also by cows, sheep, grasshoppers, snails and many other herbivores. The results are illustrated in *food webs*.

In a woodland   **In a stream**

Relationships between organisms 20

Other relationships between organisms

1 Predation Carnivores usually hunt their food and in this context are called *predators*, while the animals that they eat are called *prey*.

2 Competition This is a relationship between two different or similar organisms (*competitors*) both attempting to obtain the same food or other factor (e.g. weeds in wheat, cock robins for hens, bean seedlings for light).

3 Mutualism This is a relationship between two different organisms in which neither is harmed. In true mutualism, sometimes called *symbiosis*, each organism helps the other (e.g. lichens, where the alga provides food by photosynthesising and the fungus provides support). In other cases, sometimes called *commensalism*, it is difficult to see the advantage for one partner. *Commensals* usually have alternative food sources (e.g. sparrows on human's bird table, where the birds can obtain food easily and there is no material benefit for the human).

4 Parasitism This is a relationship between two different organisms in which one, the *parasite*, lives in or on another, called the *host* and derives food from it. (A successful parasite does not kill its host.) For example:
(*a*) flea, head louse (*ectoparasites*: living *on* host);
(*b*) tapeworms, bacteria (*endoparasites*: living *in* host).

Look at the following illustrations and decide by name which relationship they show. Check that the numbers given correspond to the names of relationships 1–4 above.

hawk and mouse **(1)**
hawks for mouse **(2)**

bee collecting nectar
+ pollinating flower **(3)**

aphid on rose **(4)**
ladybird + aphid **(1)**

seal and penguin **(2)**

stags 'rutting' for
doe **(2)**

trees growing taller
towards light **(2)**

ORGANISMS AND THE ENVIRONMENT
2.3 Diseases and defence

Parasites as pathogens

The main cause of infectious diseases in plants and animals is the entry of small parasites into the body. These *pathogens* multiply *asexually* given *organic food, moisture* and a *suitable temperature*. Some need *oxygen* and *a lack of UV light* to thrive. These conditions are provided by the bodies of most organisms. Pathogens either damage host cells directly or produce *toxins* as waste products. These upset the host's *metabolism*, causing the characteristic *symptoms* of a disease such as discolouration, spots and galls in plants, and fever, rashes and soreness in animals.

Types of pathogen

Viruses (size = 0.00001 mm) Simple protein sheath around DNA strand. *Not living cells* as they use host's energy and materials. As they multiply the host cells are killed.

Diseases caused by viruses: colds, 'flu, polio, chicken pox, rubella.

Bacteria (size = 0.001 mm) Living cells. DNA but no proper nuclei, mitochondria or endoplasmic reticulum. The wall surrounding each cell is made of *cellulose*. Found throughout the biosphere.

Diseases caused by bacteria: TB, boils, cholera, syphilis, pneumonia.

Protozoa (size = 0.1 mm) Recognisable animal cells, each with proper nucleus. Nearly always carried by *vectors* to host.

Diseases caused by protozoa: sleeping sickness, malaria, dysentery.

Fungi (size = very variable) Surrounding wall made of chitin. Branching forms have no cross walls. Make very small spores, so mainly spread by *wind*.

Diseases caused by fungi: rusts, smuts, mildews, thrush, Athlete's foot.

Venereal (sexually-transmitted) diseases (VD)

Gonorrhea (*pathogen = bacterium*)
Transmission: Skin contact between sex organs during intercourse or between mother's vagina and baby during birth.
Symptoms: Itching and burning sensation when passing urine. If untreated may lead to heart disease, blindness and arthritis.
Treatment: Antibiotics and sulphonamide drugs.

Syphilis (*pathogen = bacterium*)
Transmission: As with gonorrhea but also through skin scratches and across placenta from infected woman to her baby.
Symptoms: Copper-coloured sore at site of infection. After 3–6 weeks, fever, body rash, sore throat, hair falling out in patches, ulcers in mouth and vulva. If untreated may damage heart, blood vessels and central nervous system.
Treatment: Antibiotics and drugs containing arsenic.

Non-specific urethritis (NSU) (*pathogen = unknown*)
This is a general description of conditions resulting in inflammation of the urethra. It is more common in males.
Transmission: Probably sexual contact.
Symptoms: Pain and burning sensation when passing urine.
Treatment: Antibiotics.

Acquired immune deficiency syndrome (AIDS) (*pathogen = virus*)
The body's immune system is broken down. Many rare infections and cancers can then attack the defenceless body, usually causing death.
Transmission: Contact with bodily fluids from an infected person.
Symptoms: Various.
Treatment: None.

Other types of disease affecting man are:
(a) Deficiency in diet, e.g. scurvy (lack of Vitamin C) – curable;
(b) Inherited/genetic, e.g. Down's syndrome (monogolism) – incurable;
(c) Shortage of hormones, e.g. Diabetes (lack of insulin) – controllable.

Diseases and defence

Methods of entry into the human body

Entry of pathogens is simple, as the organisms concerned are very small. (In plants they invade through *stomates, lenticels* or *damaged tissue*, or are injected by the mouth parts of grazing animals, usually insects).

Via nose and mouth
(a) Droplets from infected people (e.g. coughs and sneezes spread colds).
(b) Food and drink can carry micro-organisms (e.g. food poisoning).

Via blood transfusions
From infected donors (e.g. AIDS and other viral infections).

Via semen
Sexual intercourse (e.g. AIDS and other viral infections).

Via skin
(a) Direct penetration by micro-organisms (e.g. Bilharzia).
(b) Through bites (e.g. fleas and bubonic plague, dogs and rabies).
(c) Through cuts and abrasions (e.g. general bacterial infection).
(d) Indirect contact with infected people via their belongings, such as towels or cutlery (e.g. Athletes foot, hepatitis).
(e) Direct contact of skin (e.g. chicken pox, VD, measles).

The body's lines of defence

There are two lines of defence, the first is *external* and aims at preventing entry of micro-organisms. The second is *internal* and copes with micro-organisms that have crossed the first line.

First line of defence
(a) *Tears* are antiseptic.
(b) *Skin:* keratin forms a waterproof barrier. Sebum + sweat are antiseptic.
(c) *Mucous membranes* inside the nose, mouth, trachea and oesophagus are inpenetrable to pathogens.
(d) *Cilia* in air passages sweep particles, including pathogens, to the throat for swallowing.
(e) *Blood-clotting:* bleeding flushes out some pathogens; clot and scab provide a temporary barrier.
(f) *stomach acid* kills bacteria in food.

Second line of defence
(a) *Polymorphs* ingest pathogens at sites of infection.
(b) *Lymphocytes* produce antibodies to neutralise toxins made by the pathogen.
(c) *Interferon:* a protein secreted by individual cells, inhibiting viral reproduction.

Artificial defences

Vaccination There are two methods:

Passive vaccination	Active vaccination
Inject antibodies made by another animal, e.g. horse.	Inject dead or attenuated (weakened) pathogen.
Swift, short-lived action	Slower, long-lived action

Edward Jenner (1749–1823) noticed that milkmaids who had suffered from cowpox, a mild disease, did not contract smallpox, a disfiguring disease which sometimes killed. He used cowpox pathogen to infect his son in order to protect him from smallpox, thereby practising the first *vaccination* or *immunisation*.

Louis Pasteur (1822–1895) continued major work of economic importance on vaccines and sterile techniques.

Disinfectants (e.g. bleach in toilet) Synthetic chemicals are used on objects and surfaces to kill micro-organisms.

Antiseptics (e.g. TCP on grazed skin) Synthetic chemicals are used on humans in concentrations that kill micro-organisms but not human cells. Antiseptic surgery was pioneered by *Joseph Lister* (1827–1912). He used carbolic acid. Nowadays, *aseptic surgery* is practised; all equipment and surroundings are sterilised so that the presence of microbes is minimised.

Drugs (e.g. aspirin for 'flu and headaches) Synthetic chemicals tend to deal with *symptoms* and not the causes, which are micro-organisms.

Antibiotics (e.g. penicillin) Chemical secreted by fungi or bacteria compete with and kill pathogenic micro-organisms. Penicillin was discovered by *Alexander Fleming* (1881–1955).

Hygiene The practice of hygiene includes *washing* the body, clothes, dishes, etc., with *detergent* or *soap*. Methods of *water treatment* have been developed and include filtration and chlorination to remove suspended particles and bacteria respectively. *Sewage treatment* returns cleaned waste to our rivers.

Food preservation Many methods prevent *spoilage* (rotting). The numbers of micro-organisms are minimised by depriving them of one or more of the factors needed for reproduction or by killing them outright (e.g. deep-freezing, refrigeration, pickling, salting, jam-making, vacuum-packing, sterilisation, bottling, smoking and freeze-drying).

ORGANISMS AND THE ENVIRONMENT
2.4 Cycles of matter

Our world only holds a limited amount of the different chemicals essential to life. Instead of these chemicals running out, they are *recycled*.

Recycling links all living organisms. It involves *scavengers* and *decomposers* that feed on dead organic matter (corpses and faeces). Scavengers eat it (e.g. earthworms, pond snails); decomposers release enzymes that digest it externally before absorption and are also called *saprophytes* (e.g. bacteria). They cause decay and release carbon dioxide and minerals needed for plant growth. (If added to the end of a food chain, the chain ends can be connected, becoming cycles.)

By contrast, *energy is not recycled*. It flows through an ecosystem as summarised below.

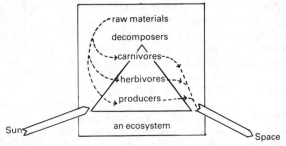

The carbon cycle

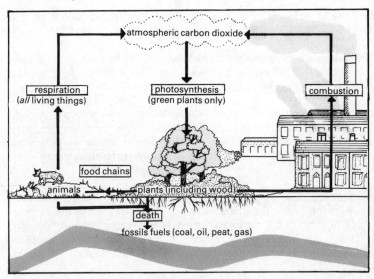

The nitrogen cycle

Nitrogen is needed to make amino acids, which are used to make proteins. It is therefore essential to growth and hence life.

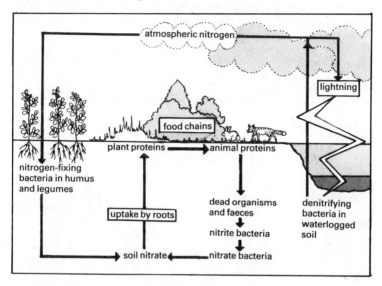

The water cycle

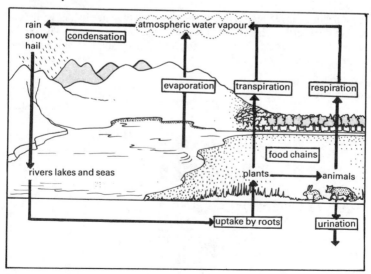

ORGANISMS AND THE ENVIRONMENT
2.5 Soil

Soil is important because it acts as a store for chemicals. Urine, faeces and dead organisms are broken down into their constituent chemicals by the soil organisms that feed on them. Plants can use the rest and pass them on to animals in food chains.

What is soil?

Soil is the topmost layer of the Earth's crust. It consists of: (*a*) rock particles; (*b*) mineral salts; (*c*) humus; (*d*) air; (*e*) water; and (*f*) organisms.

Profile of the soil and how it is formed

Frost
(cracks rock)

Acidic rain
(dissolves rock)

Roots
(widen cracks)

Mechanical
Breakdown
(grinds
particles)

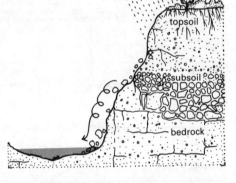

Topsoil can be investigated by shaking it well with water in a large jar and leaving it to settle. This is called a *sedimentation experiment*. The largest particles settle out first. A good soil for cultivating plants contains a mixture of particles of all sizes.

Results of a sedimentation experiment

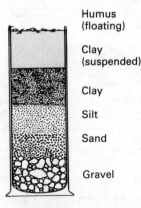

Humus
(floating)

Clay
(suspended)

Clay

Silt

Sand

Gravel

Soil 28

Types of soil	Clay	Loam	Sand
Appearance			
Structure	Small spaces between small particles (0.002 mm).	Range of particle sizes and pieces of humus.	Large spaces between large particles (0.02–2.0 mm).
Properties	Sticky and heavy when wet. Difficult to dig or plough. Stays waterlogged so no oxygen for roots. Cracks when dry.	Holds some water and oxygen. Easy to dig and plough. Ideal for plant growth.	Drains too easily. Water takes dissolved chemicals needed for plant growth away with it (*leeching*).
Treatment for cultivation	Add humus or lime to *flocculate* clay particles (i.e. coarsen the texture of the soil).	None needed.	Water constantly (*irrigate*) or add humus ('*mulch*').

Testing soil

1 Finding soil organisms

Micro-organisms (e.g. fungi, bacteria): use agar plates

(*a*) Lift the lid as little as possible;
(*b*) Use clean wire loop to scatter soil on agar surface;
(*c*) Label dish and seal with sticky tape;
(*d*) Incubate at room temperature for 72 hours;
(*e*) Any fungi or bacteria present will appear as colonies on agar surface.

Small animals (e.g. beetles, worms): use Tullgren funnel

(*a*) Shine 40 watt bulb over funnel;
(*b*) Leave for 5 minutes;
(*c*) Place beaker of 70% ethanol under funnel for collecting dead specimens.

Larger animals (e.g. moles, shrews, mice) can be captured in 'pitfall traps' or 'small mammal traps'.

29 Soil

2 How quickly water drains through soil

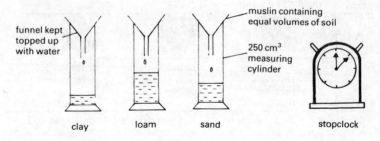

(a) Set up the apparatus as in diagram;
(b) Start stopclock and fill the funnels at the same time;
(c) Measure how much water passes through each sample in one minute.

3 Water content

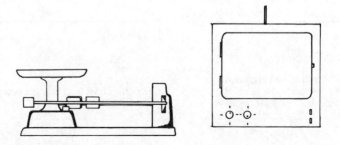

(a) Carefully 'weigh' a soil sample
 (remember to take account of its container);
(b) Place it in an oven at 100°C for at least 1 hour
 (this makes soil water evaporate);
(c) 'Reweigh' sample;
(d) Calculate *percentage water in soil* =
$$\frac{\text{mass of wet soil} - \text{mass of dry soil}}{\text{mass of wet soil}} \times 100\%,$$

Soil 30

4 Humus content (using soil sample from 3 above)
(a) Take the dry soil sample;
(b) Place it in a crucible over a hot Bunsen flame (this burns away humus, leaving only minerals);
(c) 'Reweigh' sample;
(d) Calculate *Percentage humus in soil* =
$$\frac{\text{mass of dry soil} - \text{mass of burnt soil}}{\text{mass of wet soil}} \times 100\%.$$

5 pH of soil
(a) Place 2 cm depth of soil in a test tube;
(b) Add 2 cm of barium sulphate powder;
(c) Add 5 cm depth of distilled water;
(d) Add 5 drops of soil indicator solution;
(e) Shake well and leave to settle;
(f) Compare the clear liquid on top against the colour chart.

Likely results:

acid	4	5	6	7	8	9	alkaline
	peat		most soils		chalk		

6 Air content

(a) Remove a small tin of water from a glass trough and measure its volume;
(b) Discard the water and push the tin into fresh soil, then dig it out;
(c) Put soil in trough;
(d) Using a measuring cylinder, bring the level back up to the original level mark;
(e) Calculate *percentage air in soil* =
$$\frac{\text{volume of water added in } (d)}{\text{volume of tin can from } (a)} \times 100\%$$

ORGANISMS AND THE ENVIRONMENT
2.6 Practical ecological study

Records for an area need to be built up over many years to give an accurate picture and an understanding of the changing balance within it. Practical studies in ecology are not easy!

The first steps are to:

(a) Define your study area;
(b) Find out what species are present (keys and identification books will help);
(c) Find out how many of each species there are (this is where the different techniques come in).

It is impossible to do this thoroughly for even a small area, so we take a *sample*. This is usually 10% of an area. Many sampling methods have been devised but there are always problems, e.g.

(a) Plants are more noticeable if they have brightly coloured flowers, large or variegated leaves or tall upright stems;
(b) Animals are more noticeable if they are coloured for courtship display;
(c) Well-camouflaged animals may be overlooked altogether!

Sampling methods

As a general rule, plants stay still, but animals do not! There are therefore different ways of sampling them and these methods dictate what measurement is made.

For plants
(a) *Point quadrats*
Needles are lowered in a frame and the number of times a needle hits a species recorded. Measurement made = frequency

(b) Square quadrats

There are usually wire squares with sides 0.5 m long and can be arranged as follows:

(i) Grids

Quadrats are arranged at random in an area by throwing them with eyes shut (be careful!) or using random numbers to place them on axes. This method gives a general picture of an area.

(ii) Belt transects

Quadrats are placed side by side along a tape or string stretched across the study area. This method can show the effects of gradients on environmental factors within an area.

Measurements made within square quadrats are usually *density* or *cover*.

Density

0.5 m

0.5 m

In this quadrat the plant has a density of
6 per 0.25 m² ≏ 24 per m²

Cover

0.5 m

0.5 m

In this quadrat the plant has a cover value of about 25%. (Remember cover values can be over 100%, e.g. where grass grows under trees)

Both plants and animals can simply be *counted* or their *biomass* measured. Keeping *clear records* is important.

33 *Practical ecological study*

For animals

(a) Observation and counts
 Helped by e.g.:
 'Hides'
 Binoculars
 Microscopes
 Tally counters
 Remotely-controlled cameras
 Infra-red film (for night-time)

(b) Capture and Recapture
 Techniques
 Helped by e.g.:
 Pitfall traps
 Water traps
 Water net sweeps
 Netting and 'ringing' birds
 Marking with non-toxic paint
 Butterfly nets

Environmental factors Factors that can be measured and may influence the occurrence and distribution of a species are:

Temperature
Drainage
Water content of the soil
Humus content of the soil
Air content of the soil
Air humidity
pH
Light intensity
Wavelength (colour) of light

Daylength
Oxygen concentration
Carbon dioxide concentration
Turbidity (cloudiness) of water
Minerals present
Poisons present
Shelter
Water flow/current

Environmental probes can measure many factors or separate tests can be carried out for each.

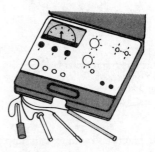

Make sure that you are familiar with *one habitat* that you have studied and can:

(a) Name the habitat (e.g. playing field, freshwater pond);
(b) Name at least five species living there;
(c) Outline at least one relationship between the named species;
(d) Describe how to test for one environmental factor;
(e) Suggest how this environmental factor influences the species present.

Presenting findings

Apart from illustrations and written accounts, there are two common ways of presenting the data from line transects.

(a) *Histograms (bar charts)*

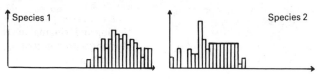

(b) *Kite diagrams*

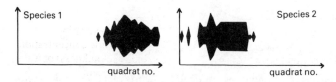

These can be linked to environmental factors and help to show relationships between organisms and their surroundings, for example:

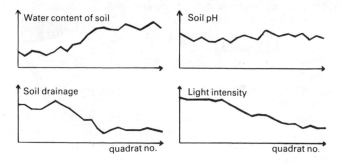

The results above show a line transect starting in an open meadow and running into a woodland stream, recording the distribution of plant species. Look for relationships. For example, species 1 is found only in very wet soil (it could be a moss or a rush), species 2 is found only in open meadow (it could be a flower such as Meadow Sweet).

(c) *Food chains, webs* and *pyramids* can be constructed after counting, measurement of biomass, observation and reference to text books.

ORGANISMS AND THE ENVIRONMENT
2.7 Population size and control

Population growth curve

Population studies are easily carried out on micro-organisms in the laboratory. Growing bacteria or yeast cells in a suitable medium reveals a pattern of growth shared by many other species. It follows an *S-shaped* or *sigmoid* growth curve. *Four stages* are recognised.

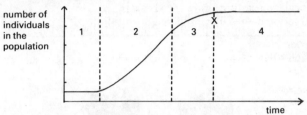

1. *Lag phase* The population is becoming established. Growth is slow but numbers are accelerating.
2. *Log phase* Conditions are ideal. Growth is very rapid. It is logarithmic (hence log phase).
3. *Limiting phase* Environmental factor(s) begin to limit growth rate, increasing death rate or decreasing birth rate. Numbers still increase, but more and more slowly until growth stops at X. In natural situations, factors that **stop** log growth are:
   Shortage of food, water, light or shelter;
   Toxic wastes accumulating;
   Overcrowding, leading to aggression and spread of disease, and
   Predators.
   Competition between individuals comes into play when conditions are no longer ideal.
4. *Stationary phase* At X, birth rate = death rate, and there is no overall change in population numbers thereafter. When the curve reaches this level, the environment is *saturated* and can support no more individuals of that species.

Human population size

In the UK
Population size depends on three factors:

(*a*) migration and emigration (people leaving and entering the country);
(*b*) birth rate;
(*c*) death rate.

The first factor has very little effect compared with the other two. Birth rate has recently dropped, due to the introduction of more effective birth control methods and family planning. Death rate has been slowed because of advances in our health care. There is a general *ageing* of the population. This could lead to a **crisis** in our health services, due to:

Contraceptives reducing birth rate,
Reduced death rate has been effected by
Improved water supply and
Sewage treatment
Inoculation, antibiotics and other drugs
Surgery and other medical skills.

Worldwide

The graph below shows the growth in the human race over the past 2000 years:

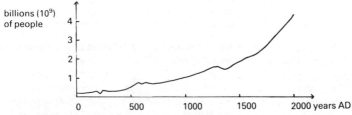

Man's population growth curve is in its log phase.

Man is different from other species in that he can control his fate. If, as seems likely, man's growth continues to follow the typical curve, he needs to plan ahead now, before reaching the limiting phase. Man should limit his own growth and not make too many demands on his environment if he is to achieve an acceptable stationary phase. To **correct** the current trends he must consider:

Contraception;
Overfishing, tree felling, and using renewable resources too fast;
Recycling minerals and *not* draining non-renewable resources;
Recreational needs in overcrowded areas;
Energy sources such as solar, tidal and wind power;
Control of pollution;
Technology in food production.

Already food is short in some countries. People either get too little to eat and suffer *starvation* (*marasmus*), or they do not get the right types of food, and suffer *malnutrition*. Malnutrition leads to *deficiency diseases* (e.g. *kwashiorkor* caused by lack of protein and scurvy caused by lack of vitamin C).

The countries of the *Third World* are usually worst affected. There, *climate* is least favourable, *infectious diseases* go largely untreated and *political situations* are more unstable. It is easy to follow what are sometimes called the *vicious circles of poverty*. Examples are illustrated below:

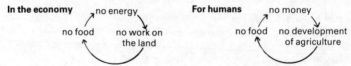

Even when food supplies are available, it is often difficult to ensure that the food is *distributed* to those who need it.

Methods of contraception

Method	What it is	How it works
Rhythm method ('safe period')	No sexual intercourse around the time of ovulation. Calculated by taking temperature daily and referring to chart.	Sperms are not introduced at times when fertile ova are in the oviduct. Calculations are difficult. *Method is unreliable.*
Withdrawal (coitus interruptus)	Withdrawal of the penis from the vagina before ejaculation.	Supposedly prevents sperms reaching ovum. *Very unreliable.*
Sheath, condom, Durex, French letter	A thin rubber sheath is unrolled on to the erect penis before sexual contact.	Forms a barrier preventing sperms entering the vagina.
Diaphragm Dutch Cap	Placed over the cervix prior to intercourse. Removed several hours later.	Forms a barrier preventing sperms reaching the ovum.
Spermicides – creams/aerosols	Sperm-killing chemicals are put near the cervix (or used with a sheath) before intercourse.	Sperms are killed before they reach the ovum.
Contraceptive pill	Female hormones are taken orally each day (occasionally similar hormones are injected).	Chemicals simulate pregnancy, stopping ovulation and plugging the cervix with mucus.
Sterilisation (vasectomy in male)	Vas deferens is tied and cut (local anaesthetic). Oviducts are blocked or cut (general anaesthetic).	Operation completely stops sperms or ova reaching appropriate fertilisation sites.
Intra Uterine Device (IUD) – loop, copper 7 coil	Shaped plastic or copper placed into uterus by doctor.	Foreign body in uterus prevents implantation.

2 ORGANISMS AND THE ENVIRONMENT
2.8 Man and the environment

Agriculture

The world's farmers need to increase the amount of food that they produce to keep up with the ever-increasing population. To grow more food is not just a question of sowing more seeds; there may not be enough room! It means growing bigger, better plants with less waste.

Some agricultural methods have had their problems, so farmers have to consider how their actions may affect the ecosystem. They also have to work out *cost-effectiveness*, i.e. whether the cost of increasing production is worthwhile in cash return for their harvest.

Consider the following agricultural practices. They aim to maintain topsoil, increase soil fertility, improve stock and minimise competition. They improve *crop yield* so we have more plants to eat and to feed to our animals. Remember *crop rotation* and **ploughing** (the letters of 'ploughing' will help you to recall the most important methods).

Crop rotation: Different crops grown on land in successive years allowing a legume to restore nitrogen to the soil, at intervals.

Ploughing: this breaks up soil clumps and aerates it ready for sowing. It also buries stubble.

Liming: this causes clay to flocculate, improving drainage and aeration. It also neutralises acid soils.

Organic and inorganic fertilisers: these include manure and compost which provide nutrients and humus to improve soil texture and NPK (nitrogen, phosphorus, potassium) fertilisers which artificially add nutrients.

Use of pest control: *chemical* toxins get rid of pests (e.g. insecticides kill insects, herbicides kill weeds). *Biological* control uses a predator or parasite to get rid of a pest (e.g. ladybirds eat greenfly).

Growth hormones: these are made artificially and accelerate plant growth (e.g. auxins).

Hedging and terracing: these prevent soil erosion by wind and water.

Irrigation: this provides essential water for dry soils.

New varieties: plant breeders select useful features (e.g. large fruits).

Greenhouses: these can provide a warmer environment with greater carbon dioxide concentration.

Even with all this knowledge, *famine* is common in some parts of the world, while others have so much *surplus* that food is stored in butter and grain 'mountains'. The reasons for this imbalance stem from politics and economics, as well as distribution problems.

Pollution

Pollution is the addition of substances or energy to the environment which upset the normal balance in the biosphere. Pollution can usually be prevented. Often it is not, because the cost or effort is too great. If you remember that there should be a **law** against it, the letters of the word 'law' will remind you of the main areas that pollution affects, i.e. **l**and, **a**ir and **w**ater. Some important examples are given below.

	Pollutant	Source(s)	Effect(s)	Prevention
Land	Litter	Human carelessness	Unsightliness; death of animals	Education; better clearing services
	Insecticides	Control of disease vectors; crop protection	May concentrate along food chains, killing consumers	Legislation (UK banned DDT that killed falcons)
	Radio-activity	Nuclear reactor waste dumping; accidents; bombs	Mutations	Better waste silos; reduced use
Air	Soot (C)	Factories; coal-burning; steam engines	Smog coats leaves, reducing photosynthesis	Clean Air Act 1952; smokeless fuels; chimney filters
	Carbon monoxide (CO)	Car exhaust fumes	Replaces O_2 in mammals' blood	'After-burners' in cars
	Sulphur dioxide (SO_2)	Smelting; coal burning	Acid rain; smog; bronchitis	SO_2 extraction units; smokeless fuels
	Chlorofluoro methanes	Aerosol propellants (spray cans)	UV light penetration breaks down ozone layer in Earth's atmosphere	Not yet invented
	Noise	Aircraft; traffic	Damage to hearing; stress	Legislation; noise insulation; double glazing
Water	Sewage	Human	Kills fish because thriving bacteria use up all O_2	More sewage treatment works
	Artificial fertilisers	Excessive use by farmers	Kills fish because thriving bacteria use up all O_2	Restricted use
	Oil	Tanker washing at sea; spillages	Oil kills birds and contaminates beaches	Legislation; accident prevention
	Mercury	Fungicides; chemical works	Paralysis in humans	Effluent purification
	Heat	Power stations	Thriving algae clog rivers	Cooling units

Conservation

Conservation is our attempt to reduce the effects that we have on the biosphere. These are the results of land usage, farming methods, overhunting, overfishing, mining, felling trees and pollution. Conservation means taking care of our environment so that it is a fit place for all living things, including ourselves. (Recreation and enjoyment are essential to man's health.) It also means taking care of plants and animals so that they do not become extinct. Genes from wild plants and old, rare breeds of farm animals can now be used by genetic engineers to help breed better varieties.

We have not always foreseen the consequences of our activities, e.g. the killing of birds of prey by the introduction of DDT and the *greenhouse effect* (the rise in global temperature as CO_2 levels increase in the atmosphere after tropical rain forests are felled). We need to think. Living resources will *recycle* themselves if given sufficient time. Harvesting should not exceed replacement rate, i.e. reproduction rate. Non-renewable resources (e.g. minerals, oil and gas) could run out altogether.

What can be done?

Individuals can consider:

(a) litter and paper collection, for re-cycling;
(b) use of 'bottle banks' to re-cycle glass;
(c) observing the Countryside Code;
(d) refusing to buy fur coats, skin handbags, ivory, cosmetics;
(e) joining a campaign organisation, e.g. 'Save the Whale';
(f) refusing to buy articles with over-elaborate packaging.

Organisations and *politicians* can consider:

(a) legislation (laws) to ban effluents, protect species, etc.;
(b) limits on hunting and fishing seasons, net size etc.;
(c) establishing more amenity areas, e.g. Nature Reserves and National Parks;
(d) afforestation (tree planting) schemes;
(e) careful planning of sites for industry, roads, reservoirs, etc.;
(f) new (acceptable) energy sources, e.g. tidal, wind and solar power;
(g) new food sources, e.g. soya meat substitutes and single-cell proteins;
(h) use of fibre glass and carbon fibre plastics instead of metals;
(i) prevention of soil erosion by hedge-planting and contouring;
(j) education;
(k) recycling paper, metal, glass, etc.

ORGANISMS AND THE ENVIRONMENT
Sample questions

2

1. The total energy entering a food chain is that present in
   A sunlight   B the producer
   C all the organisms   D the primary consumer

2. The world human population by 2000 AD is likely to be
   A 2000 million B 6000 million C 10 000 million D 100 000 million

3. A food chain contains 1. mice 2. flea 3. wheat 4. cat. Place these organisms in order, starting with the producer.
   A 3 1 4 2   B 2 1 4 3   C 2 4 1 3   D 3 2 1 4

4. Which of the following organisms converts nitrogen gas to nitrate?
   A nitrifying bacteria   B nitrogen-fixing bacteria
   C decay fungi and bacteria   D denitrifying bacteria

5. The amount of energy passed from herbivores to carnivores is about
   A one half   B one quarter   C one tenth   D one hundredth

6. Who first used immunisation against smallpox?
   A Lister   B Fleming   C Pasteur   D Jenner

7. Defences against pathogens do not include
   A antigens   B stomach acid   C sweat and tears   D polymorphs

8. Water is removed from the water cycle through
   A perspiration   B photosynthesis
   C transpiration   D condensation as rain

9. The term 'humus' refers to the presence in the soil of
   A organisms   B leaf litter   C dead matter   D peat

10. In a study of the environmental factors *directly* affecting plants one need not consider
    A herbivores   B mineral salts   C carnivores   D temperature

11. The layer of the Earth's surface that supports the biotic (living) component is called the
    A ecosystem   B crust   C habitat   D biosphere

12. The best soil for plant cultivation is
    A loam   B sand   C clay   D humus

13. Which of the following is *not* recycled in an ecosystem?
    A carbon   B energy   C water   D nitrogen

14. In ecological studies, the number of individuals in a unit area is called
    A cover   B biomass   C density   D quadrat

15. A flea living on a cat is an example of
    A ectoparasitism   B symbiosis
    C endoparasitism   D predation

16 (*a*) Describe *briefly* what is meant by the term *amenity* area. (2)

(*b*) Wye Council reduced the height of spoil heaps near a town, planted grass, shrubs and trees and set up a picnic area. Tay Council did nothing to the spoil heaps but they put a fence around them to keep people away from possible danger.

  (*i*) What effect does Wye Council's efforts have on the town's people? (1)
  (*ii*) What effect does Wye Council's effort have on wildlife? (1)
  (*iii*) State *one* advantage of Tay Council's action. (1)
  (*iv*) State *one* disadvantage of Tay Council's action. (1)

(*c*) The activities of man have led to some plants and animals becoming extinct in the world.

  (*i*) Name one species of plant or animal known to you that is in danger of becoming extinct. (1)
  (*ii*) State one reason why this species is rare. (1)
  (*iii*) Suggest what could be done to save this species. (2)

17 People with chronic bronchitis suffer from a persistent cough and shortness of breath. The graphs below show the percentages of smokers and non-smokers, aged between 35 and 69, who suffer from chronic bronchitis in two towns, **A** and **B**. **A** is an industrial town. **B** is a small sea-side town. Study the graphs below.

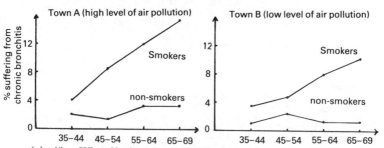

(*a*) (*i*) What % of smokers aged 55–64 in **A** have bronchitis? (2)
  (*ii*) What % of smokers aged 35–44 in **B** have bronchitis? (2)
(*b*) What evidence is there that smoking is more important than air pollution in causing chronic bronchitis? (2)
(*c*) What evidence is there that air pollution is one factor which increases the chance of suffering from chronic bronchitis? (1)
(*d*) Name *three* other effects that heavy smoking has on the human breathing system. (3)

18 (*a*) How have man's activities affected his environment? (10)
(*b*) Describe human population growth over the last 2,000 years. (2)
(*c*) Outline how man can control his population growth. (8)

THE INDIVIDUAL ORGANISM
3.1 Studying cells

Cell structure

Typical plant and animal cells are illustrated. The features that they have *in common* are labelled on the left-hand side and those that are *different* on the right-hand side.

A typical plant cell (e.g. moss leaf cell)

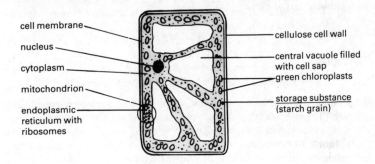

A typical animal cell (e.g. human cheek cell)

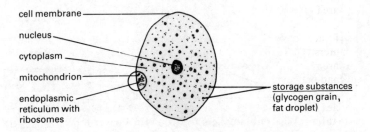

(Remember that you cannot see mitochondria and endoplasmic reticulum using the light microscopes in schools – see p. 45.)

Cell organelles (components)

In the list below, a (*P*) after a description reminds you that the organelle is found in most plant cells and an (*A*) that it is found in most animal cells.

Cellulose cell walls give plant cells shape and strength. Cellulose is freely permeable to all molecules. (*P*)

Cell membranes are true cell boundaries. They are made of lipids (fats) and proteins. They are permeable to water and gases, but selectively permeable to larger molecules. (*P*) (*A*)

Cytoplasm is a jelly-like matrix. (*P*) (*A*)

Chloroplasts are the sites of photosynthesis, trapping sunlight energy in green chlorophyll and converting it into chemical energy. (*P*)

Mitochondria are the sites of respiration, providing the energy needed by the cell. (They are similar in size to chloroplasts.) (*P*) (*A*)

Endoplasmic reticulum is a network of fine tubes and flattened sacs extending throughout the cytoplasm, organising enzymes and helping with transport across the cytoplasm. (*P*) (*A*)

Ribosomes are very small particles responsible for assembling proteins. They are found mostly on the endoplasmic reticulum. (*P*) (*A*)

Nuclei are active central banks of information and are often likened to computers. They communicate with the cytoplasm using DNA and RNA. They are responsible for cells' day-to-day activities and for passing genetic information to offspring. (*P*) (*A*)

Vacuoles are liquid-filled spaces whose functions include food storage, waste disposal and pushing the cytoplasm against the cell wall to give the cell turgor. (*P*) (*Note:* Some animal cells have *temporary* contractile vacuoles which fill with water and then discharge their contents outside the cell membrane.)

The light microscope

The invention of the light microscope is usually attributed to *Anton van Leeuwenhoek*. Check that you can label the diagram below.

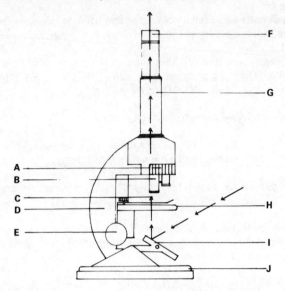

A	turning turret	F	eyepiece containing eye lens
B	nose pieces containing objective lenses	G	ocular tube
C	path of light	H	stage
D	body	I	mirror
E	focusing wheel	J	base

Magnification Total magnification that you see when you look at a prepared microscope slide is calculated by *multiplying* the strength of the eye lens by the strength of the objective lens, for example:

Eye lens	Objective lens	Total magnification
× 4	× 10	× 40
× 10	× 100	× 1000

The very best light microscopes can magnify about 1500 times.

Other types of microscope Electron microscopes use an electron beam instead of light. They can magnify about 500 000 times. The transmission electron microscope (TEM) examines thin sections, and the scanning electron microscope (SEM) views whole small specimens in 3-D.

Studying cells 46

Cells in perspective

Cells of different living things vary greatly in size. Most cells are between 0.001 mm and 0.02 mm. However, they can be as big as the yolk of an ostrich egg (75 mm), as long as an elephant's nerve cell (2000 mm) or as small as a boil bacterium (0.0001 mm). The table below places cells in perspective, starting with the smallest units and building up through the recognised levels of organisation.

Level of organisation	Examples
atoms	O H
simple inorganic molecules	O_2 H_2O
complex organic molecules	proteins
cells	cheek cell onion cell
tissues (groups of similar cells carrying out a certain function)	epithelium palisade mesophyll
organs (made up of different tissues, each of which contributes to the whole)	stomach leaf
organ systems (made up of organs working in conjunction)	alimentary canal branches of leaves
organisms (whole bodies)	*Homo sapiens* (human) 60,000,000,000,000 cells *Quercus robur* (oak)

We then consider (*a*) *breeding pairs*, the basic unit of sexual reproduction, (*b*) *populations*, groups of similar organisms found in a specified area, and (*c*) *species*, a number of populations from different areas.

THE INDIVIDUAL ORGANISM
3.2 Specialisation

Cells

Identify the features listed below each diagram. These show how the six cell types are *adapted* or *specialised* to their functions.

Red blood cell

concave shape gives larger surface area for O₂ diffusion
haemoglobin to carry O₂
no nucleus

Polymorph (white blood cell)

changing shape to engulf bacteria
enzymes to digest bacteria
lobed nucleus

Sperm

acrosome to penetrate ovum
little cytoplasm to keep mass small
tail for swimming
mitochondria for energy

Guard cells in leaf

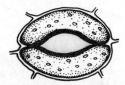

thick cell wall on one side to effect bending and opening of stoma
chloroplasts present

Neuron (nerve cell)

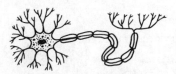

long shape to reach body parts
dendrites to connect with other neurons
sheath to speed up transmission of nervous impulse

Windpipe cells

cilia to sweep dust and germs to throat for swallowing

Organs

Specialisation is often complex and even small portions of organs may show enormous cell differences and *division of labour*. Modifications in cell structure reflect the job (*function*) that each does.

Leaf

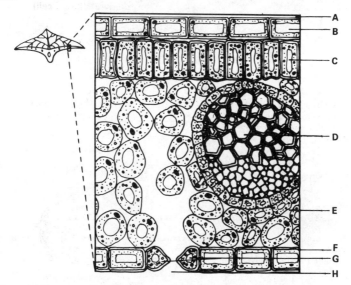

- **A** cuticle (protects and limits loss of water needed for photosynthesis)
- **B** upper epidermis (protects and supports, secretes cuticle)
- **C** palisade mesophyll near upper surface and containing many chloroplasts (receives light for photosynthesis)
- **D** vascular bundle of xylem, phloem and cambium (transports water and food to and from leaf)
- **E** spongy mesophyll (allows gaseous exchange)
- **F** lower epidermis (protects and supports)
- **G** guard cells (control opening and closing of stomata)
- **H** stoma concentration on lower surface (limits loss of water needed for photosynthesis)

Note: Other ways in which leaves receive maximum light are:

(a) Leaf shape provides a large surface area.
(b) Leaves are held at right angles to light source.
(c) The arrangement of leaves on branches avoids overlapping. This is called *leaf mosaic*.

Specialisation

Part of mammal's skin

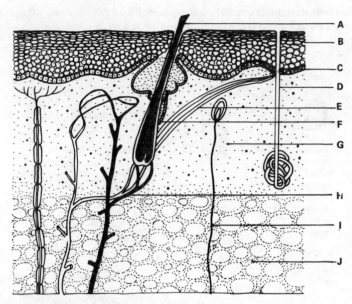

- **A** hair (forms insulating layer, whiskers warn of danger – e.g. cat entering confined space)
- **B** cornified epithelium (water- and bacteria-proofs)
- **C** malpighian layer (renews cells)
- **D** sweat pore + gland (excrete)
- **E** hair erector muscle (controls temperature)
- **F** sebaceous gland (waterproofs and keeps supple)
- **G** melanin-making cell (camouflages and protects from UV light)
- **H** blood capillaries (transport and temperature control)
- **I** nerve cell (detects environmental change)
- **J** fat (insulates against cold, produces vitamin D)

The ABC of skin functions

Acts as a *sense organ* and a *homeostatic organ*. It helps also as a
Barrier against water, cuts, pathogens and UV light (melanin).
Camouflage: pigments may help organism to blend with its surroundings.
D producer: produces Vitamin D by action of sunlight on fat.
Excretory organ: gets rid of urea in *sweat*.
Fighting aid: parts can be modified (e.g. nails, hooves, claws).

Organisms

Whole organisms are specialised for a particular way of life in a particular habitat. Adaptations are listed below each example.

Roach

internal gills for gas exchange under water.
streamlined shape for swimming
fins for steering and braking
lateral line to detect vibrations
scales to avoid uptake of water

Bladder Wrack (seaweed)

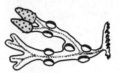

holdfast to attach to rocks
pigments for photosynthesis
air bladders to increase buoyancy and hold seaweed near the surface in light

Gnat Larva

breathing tube for gas exchange above water level
hairs to aid swimming
mandibles/jaw for meat eating

Frog

eyes allowing viewing above water level while body submerged
moist skin to help gas exchange
back legs powerful for jumping
webbed feet for swimming
front legs short for absorbing shock on landing

Owl

large eyes for good night vision
colouration for camouflage
curved beak for killing prey
talons to carry/tear prey

Flea

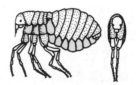

flattened shape for crawling through fur
mouthparts to pierce and suck
powerful legs for jumping

THE INDIVIDUAL ORGANISM
3.3 Cells in action

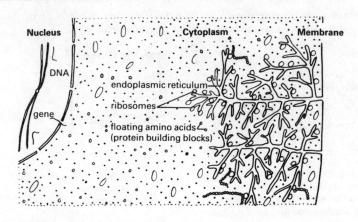

1. Gene from DNA copied into RNA
2. RNA carries 'code' for making a particular protein...
3. ...to a ribosome...
4. where amino acids are joined in a particular order to make the protein coded for.

In summary, DNA makes RNA in the nucleus. RNA makes proteins at the ribosomes in the cytoplasm. The proteins made are often *enzymes* catalysing specific chemical reactions in the cell's cytoplasm. (The structure of DNA and its role in inheritance are given on page 131.)

Chloroplast

The chloroplast is surrounded by a membrane and contains stacks of membranes that look like piles of coins. *Chlorophyll* molecules are arranged in these piles.

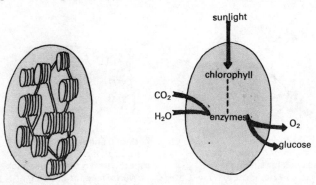

Cells in action 52

Enzymes

The properties of enzymes

(*a*) They are *proteins* made by living cells.
(*b*) They act as *biological catalysts* in chemical reactions in the body. (Catalysts are chemicals which increase the rate of chemical reactions but remain unchanged by them.)
(*c*) Each is *specific*, i.e. only converts a certain *substrate* (e.g. amylase acts only on starches).
(*d*) Each works best within a narrow temperature range, the *optimum temperature* (e.g. most human enzymes have an optimum temperature of 37°C).
(*e*) They are slowed down by a fall in temperature.
(*f*) They are *denatured* (their action is destroyed) by excessive heat, usually at temperatures above 45°C.
(*g*) Each works best within a narrow range of acid or alkaline conditions, the *optimum pH* (e.g. human pepsin has an optimum pH of 2).
(*h*) They are needed in very small amounts.

Check that you can identify the graphs describing what happens to the rate of an enzyme reaction (measured by the amount of product made in unit time) on increasing the following four factors.

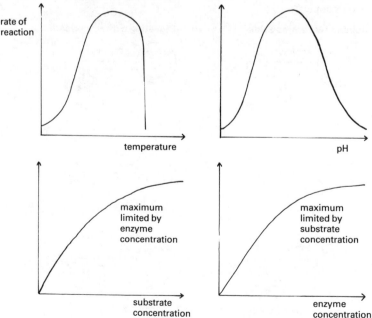

53 Cells in action

Enzymes catalyse the thousands of chemical reactions in plant and animal bodies and are *not only* involved in digestion. Even the simplest organism needs very many enzymes to carry out the seven processes that make it a living thing (p. 1).

The names of the enzymes often end in -ase, for example:

Oxidases oxidise their substrates.
Carbohydrases split up carbohydrates.
Lipases split up lipids (fats).

Enzymes are placed into two categories:

(a) *Intracellular,* which work inside cells and are more common (e.g. *catalase* that catalyses the reaction $2H_2O_2 \longrightarrow 2H_2O + O_2$ in the human liver; *starch phosphorylase* that catalyses the reaction turning starch into sugars in germinating seeds).
(b) *Extracellular,* which work outside cells and are rarer (e.g. digestive enzymes of the human gut; digestive enzymes of fungi).

How enzymes work

Proteins are complex molecules and each has a very specific shape. As enzymes only act on single or very specific substances, it is thought that enzyme and substrate(s) fit together like a lock and key, as in the following two examples:

Joining two amino acids   **Digesting a disaccharide**

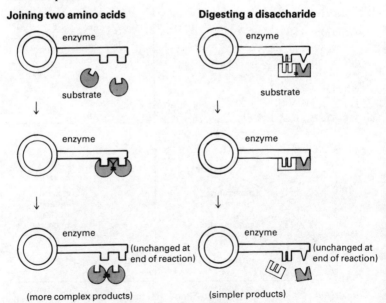

Cells in action 54

Practical work on enzymes

To demonstrate the action of saliva on starch
Method A

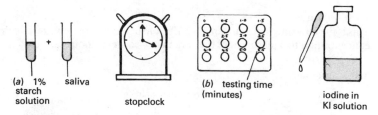

(a) 1% starch solution + saliva

stopclock

(b) testing time (minutes)

iodine in KI solution

(a) Mix the starch and saliva. At the same time start the stopclock.
(b) Place about five drops of the mixture in each dent of a labelled dropping tile.
(c) Add iodine in potassium iodide solution to the mixture to test for starch at 30 second (0.5 minute) intervals.

Results

Time in minutes	0	0.5	1.0	1.5	2.0	2.5	3.0	3.5	4.0	4.5	5.0	5.5
Colour	B	B	B	Br	Br	Br	Y	Y	Y	Y	Y	Y

B = black (starch present)   **Br** = brown   **Y** = Yellow (starch absent)

Conclusion: Saliva contains a substance that makes starch disappear. (This is, in fact, the enzyme *salivary amylase*, which converts *starch* to *maltose* and *glucose*. The activity *does not prove* this).

Method B
(a) Make two wells in a petri dish containing *starch agar*.
(b) Carefully fill one with saliva and the other with distilled water.
(c) Leave at room temperature for one hour.
(d) Flood the plate with iodine in potassium iodide KI solution.

Results

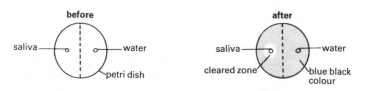

before: saliva — water — petri dish

after: saliva — water — cleared zone — blue black colour

Conclusion   Saliva contains a substance that makes starch disappear.

How substances move in and out of cells

The three main ways by which substances enter or leave cells are: *diffusion*, *osmosis* and *active transport*.

1 Diffusion

This is the movement of *gases* and *liquids* from places where they are in high concentration to a place where they are in low concentration (see below).

In the classroom

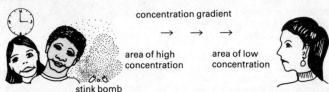

The difference in concentration between two places before diffusion occurs is called a *concentration gradient*. Diffusion results from the *random movement of molecules*. It tends to distribute molecules evenly throughout the space they occupy, so at the end of diffusion there is no longer a gradient.

In the same classroom, two hours later

no concentration gradient remains

the 'stink' has dispersed!

Diffusion, particularly in liquids, is very important in living things. It is a slow process, but can be speeded up by increasing the concentration of the substance(s) diffusing. This makes the concentration gradient 'steeper'.

Where diffusion is used for uptake of substances into cells, a large surface area of membrane is needed to supply a small volume of cytoplasm.

Consider this example:
When two lumps of sugar are put into a packet, the *volume* of sugar in a unit *doubles*. However, the *surface area* exposed *does not double* because two sides are placed together. Similarly, as an organism increases in size, i.e. volume (usually by an increase in cell number), there is a *decrease* in surface area per unit volume. Its *surface area to volume ratio* decreases.

Cells in action 56

2 Osmosis

This is a special example of diffusion. It is the movement of *water only* through a *selective* or *semi-permeable* membrane from an area where water is in high concentration to one where water is in low concentration. Osmosis tends to get rid of a concentration gradient of water.

A semi-permeable membrane is like a string shopping bag (see right) which allows small objects in and out of it, but retains objects larger than the holes. The membrane allows small molecules, but not large ones, to pass through it. *Visking tubing* is used in the laboratory as a semi-permeable membrane.

In the very highly magnified diagram of osmosis (below), sugar molecules are too big to get through the semi-permeable membrane. More water molecules will move from left to right than vice-versa.

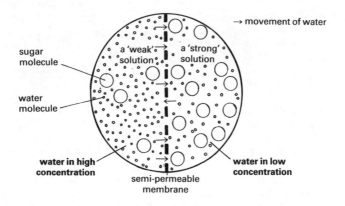

Summary of diffusion and osmosis

Diffusion If a weak sugar solution is separated from a strong sugar solution by a membrane permeable to sugar and water, both water and sugar will diffuse from areas where each is more concentrated to areas where each is weaker, until both solutions are of equal 'strength'.

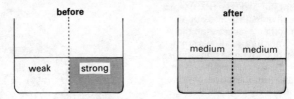

Osmosis If a weak sugar solution is separated from a strong sugar solution by a membrane permeable *only* to water, water will still diffuse from the area where it is in high concentration to an area of lower water concentration, until both areas are of equal 'strength'.

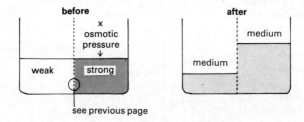

see previous page

If external pressure is applied to the strong solution before it is in contact with a weaker solution, osmosis can be prevented. In this example it would have to be applied on side X. The pressure that is just sufficient to prevent the entry of water by osmosis into a solution is known as the *osmotic pressure* of that solution.

Osmotic pressure varies according to the amount of dissolved material in a solution. The stronger a solution, the greater its osmotic pressure. However, in circumstances in which strong solutions are prevented from taking in water by osmosis, it is more accurate to describe the solution as having *osmotic potential*.

Osmosis in cells Plant and animal cells respond to osmotic flows of water in different ways. This is because plant cells have fully permeable but rigid cell walls; animals do not.

Cell type	Weaker solution	Isosmotic (equal-strength) solution	Stronger solution
Plant (e.g. onion bulb)	Cell is fully *turgid*. Cytoplasm pressed against wall.		*Plasmolysis* as vacuole shrinks and cytoplasm parts from wall.
Animal (e.g. red blood)	Cell takes in water and *ruptures*.		Cell loses water and *shrivels*.

3 Active transport

This process transports molecules across a membrane *against a concentration gradient*. It requires *energy* from respiration. Protein carrier molecules in the membranes are thought to help. Active transport can be inhibited by lack of oxygen, low temperatures or metabolic poisons like cyanide.

The table below shows examples of *diffusion, osmosis* and *active transport* in living things:

Process	Where it occurs
Diffusion	Gas exchange in lungs and tissues. Gas exchange through stomata. Glucose transport within cells. Uptake of digested food in ileum.
Osmosis	Uptake of water in root hair cells. Return of water from tissues to blood in capillaries. Opening and closing of stomata.
Active transport	Uptake of minerals in root hair cells. Aids uptake of food in ileum. Selective reabsorbtion in kidney tubules. Ion movement to generate nerve impulse. Ion movement during muscle contraction.

THE INDIVIDUAL ORGANISM
3.4 Living things need food

Nutrition refers to food and feeding. All organisms need food for:

(a) *Energy*: to maintain metabolism, move, keep warm, etc.
(b) *Making new cells:* for growth, replacement and repair.
(c) *Health:* to avoid deficiency and other diseases.

Food can be obtained in two main ways:

Autotrophic nutrition Organisms make organic food from simple inorganic substances, usually using sunlight energy (*photosynthesis*). All green plants and some bacteria practise this.

Heterotrophic nutrition Organisms obtain 'ready-made' organic food. This is practised by all consumers and can be subdivided into the categories in the following table:

Type of nutrition	Food	Used by
Holozoic	Solid organic material from the bodies of other organisms.	Herbivores, carnivores omnivores and a few specialised plants.
Saprophytic	Soluble organic material from dead or decaying organisms.	Many fungi and bacteria.
Parasitic	Organic compounds present in the body of another living organism, the host.	Some animals, flowering plants, fungi and bacteria.

The relationship between organisms with these methods of feeding is shown below:

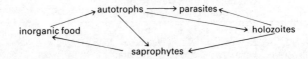

Basic nutritional requirements of living things

These can be divided into seven *food groups*:

1 Carbohydrates
2 Fats (lipids)
3 Proteins
4 Vitamins
5 Minerals
6 Roughage
7 Water

Living things need food 60

1 Carbohydrates

Units: monosaccharides (simple sugars) 6-carbon ring

Elements contained C, H and O (H : O = 2 : 1)

Types:

Monosaccharides (simple sugars) ⬡ e.g. glucose, lactose

Disaccharides (compound sugars) ⬡–⬡ e.g. sucrose, maltose

Polysaccharides (starches) ⬡–100–⬡ e.g. starch, glycogen, cellulose

All are made by green plants as they photosynthesise. Sugars are soluble in water and taste sweet. Starches are nearly insoluble in water and have no taste.

Uses: Carbohydrates provide energy; they are easily digested and respired (releasing 17 kJ per gram).

Rich sources:
(for humans)

sweets   fruit   jam   cakes   bread   potatoes   pasta

2 Fats (lipids)

Units: glycerol + fatty acids E

Elements contained: C, H and O (H : O ≈ 20 : 1)

Uses: Fats provide energy. They are more difficult to digest and are respired after carbohydrates, but are more energy-rich (releasing 39 kJ per gram). Fats are important for migrating and hibernating animals; they also help with heat insulation, buoyancy and waterproofing. They form a component of steroids and cholesterol.

Rich sources:
(for humans)

lard   butter   margarine   fried food   crisps

3 Proteins

Units: amino acids

Elements contained: C, H, O, N and often S, P.

Humans need twenty types of amino acid. Ten of these can be synthesised. The other ten, which must be eaten, are known as *essential amino acids* and the proteins containing them are known as *high grade proteins*.

Types: Amino acids are linked in long chains that can vary in
(*a*) length
(*b*) the types of amino acid contained
(*c*) the sequence of amino acids
There are thousands of types.

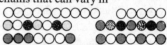

Uses: Amino acids help to build new cells. They are seldom respired. They play important roles in organisms, and are the structural base of enzymes, animal hormones (e.g. insulin), living bone, tendon and ligament, hair and skin.

Rich sources:
(for humans)

meat   fish   eggs   milk   soya

4 Vitamins

Vitamins are *organic* food substances required in very small amounts for a variety of body processes. Vitamins, with the exception of Vitamin D, are made by plants and passed to animals through food chains. Their importance to *health* can be seen when an organism is deprived of them: *deficiency diseases* arise and death may follow. Animals cannot store most vitamins for any length of time. The table on p. 62 shows human vitamin requirements and deficiency problems.

Living things need food 62

Vitamin	Source	Why it is needed	Deficiency disease
A	Milk, butter, carrots, liver	For healthy mucous membranes	Night blindness
B_1 (thiamine)	Yeast, cereals	For respiration	Beri-beri
B_2 (riboflavin)	Green vegetables, eggs	For respiration	Mouth ulcers, skin problems
B_3 (nicotinic acid)	Liver, fish, wheat	For respiration	Pellagra
B_{12} (cobalamine)	Liver	For red blood cells	Pernicious anaemia
C	Citrus fruits, blackcurrants, fresh green vegetables	For tissue repair (skin, gums and blood vessels)	Scurvy
D	Milk, butter, eggs, fish, liver, sunshine on skin	For strong bones and teeth	Rickets
E	Butter, milk, wholemeal bread	For respiration	Sterility in animals
K	Spinach, cabbage, liver	For blood clotting	Long clotting time

Early work on vitamins was done by *Frederick Gowland Hopkins* between 1906 and 1912. He used rats which he divided into two equal groups, A and B. Group A were given a balanced diet including milk, while group B were given a diet containing only carbohydrates, fats, proteins, minerals and water. After about twenty days, the diets were reversed. The results are shown below:

Gowland Hopkins experiment

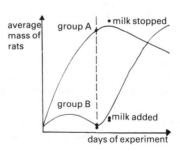

Group A showed a steady increase in mass when given milk. When it stopped they continued to grow for a short time but then lost mass.

Group B started to grow but then lost mass. When given milk they quickly increased in mass and by the end of the experiment were larger than group A rats.

Gowland Hopkins concluded that there must be 'accessary factors of the diet', provided by the small amount of milk. Today these are called *vitamins*.

5 Minerals

Minerals are *inorganic* food substances. Again, they tend to be taken in by plants and passed through food chains to animals (whose separate requirements are often different). Although needed in very small quantities, absence can cause *deficiency diseases* and even death. They are needed as:

(a) parts of various organic molecules (e.g. nitrogen, sulphur, phosphorus in proteins);
(b) parts of enzyme activators (e.g. in respiration);
(c) parts of pigments (e.g. iron in haemoglobin);
(d) structural components (e.g. calcium and phosphorus in bone);
(e) causes of osmotic differences (e.g. in kidney tubules);
(f) causes of electrical differences (e.g. in nerves and muscles).

Apparatus for studying plant mineral requirements

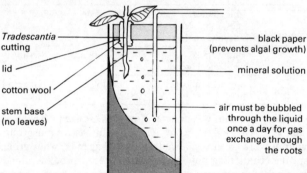

The table below shows the elements required by *plants* in mineral salts intake:

Elements	Sources	Functions
Nitrogen (N)	Nitrate ion (NO_3^-) in soil, bacteria in root nodules or artificial fertiliser	Part of proteins, RNA and DNA
Phosphorus (P)	Phosphate (PO_4^{2-}) in soil or artificial fertiliser	Part of proteins, RNA, DNA and ATP
Potassium (K)	Potassium ion (K^+) in soil or artificial fertiliser	Enzyme activator in photosynthesis and respiration
Sulphur (S)	Sulphate (SO_4^{2-}) in soil	Part of proteins, RNA and DNA
Calcium (Ca)	Calcium ion (Ca^{2+}) in soil	Part of cell walls
Iron (Fe)	Ferrous ion (Fe^{2+}) in soil	Chlorophyll synthesis
Magnesium (Mg)	Magnesium ion (Mg^{2+}) in soil	Part of chlorophyll molecule

Living things need food 64

The table below shows the elements required by *humans* in mineral salts intake:

Elements	Sources	Functions
Phosphorous (P) Calcium (Ca)	Milk and other dairy products, bread, meat, eggs	Bone and tooth formation
Iron (Fe)	Liver, spinach	Haemoglobin production in red blood cells
Potassium (K) Sodium (Na)	Table salt, vegetables	Balance essential to muscle and nerve functioning
Iodine (I)	Table salt, seafood	Production of thyroxin

6 Roughage

Roughage (fibre) is made by plants. When eaten by animals other than herbivores, it is not digested, so it passes through the body unchanged. It gives bulk to the food as it travels along the gut. In humans, it prevents *constipation* and other intestinal diseases. Roughage is found in bran products, fresh fruit and vegetables, and wholemeal bread.

7 Water

Although not strictly a food, plants and animals take in water with their food and it is critical to both for health and survival. Without water, humans suffer from *dehydration* and soon die. Water input must equal water output if water balance is to be maintained.

Water balance for humans

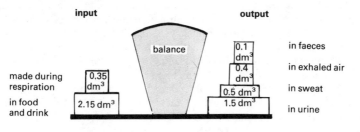

Balanced diet for humans

A balanced diet provides the *right foods* in the *right amounts* at the *right times*. The right foods include some from each of the seven food groups: carbohydrates, fats, proteins, vitamins, mineral salts, roughage and water. The right amounts depend upon five factors:

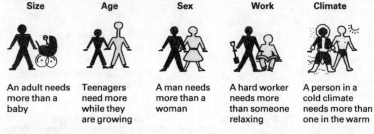

Size	Age	Sex	Work	Climate
An adult needs more than a baby	Teenagers need more while they are growing	A man needs more than a woman	A hard worker needs more than someone relaxing	A person in a cold climate needs more than one in the warm

The 'right times' usually means two or three *regular* meals that provide vitamins daily.

The energy requirements of humans (in kilojoules) are shown below:

baby	3000 kJ
toddler	5000 kJ
teenage girl	9500 kJ
teenage boy	11500 kJ
woman doing light work	9450 kJ
man doing light work	11550 kJ
man doing heavy work	15000 kJ

Special diets for humans

Not all foods are 'good' for everyone. Some people have *allergies* to a particular food (e.g. strawberries, or gluten in wheat). They suffer symptoms similar to asthma and eczema when they eat the problem food and so require a diet that avoids it. There is growing concern over allergy to *food additives* such as colourings, preservatives and flavour enhancers. An international code exists to identify them.

People who are *ill* or *convalescing* and not leading their normal active lives have special, lighter diets planned for them, while *diabetics* have to control carefully their carbohydrate intake.

Many others have a tendency to become *overweight*. These people are often unhealthy because, although they eat a lot of food, they may not get enough minerals and vitamins. Excess energy foods are stored as fat under the skin so a *slimming diet* is needed to reduce the kilojoule intake, while maintaining a balance of food groups.

Living things need food 66

Outline of most common food tests

Method and reagent(s)	Heat	Colour changes
Testing for sugar food sample extract with water — (a) add 1 cm depth *Benedict's solution* for simple sugars (b) add a few drops of *dilute HCl* and repeat (a) for compound sugars	+	sugar absent: blue sugar present: brick red (green if very little)
Testing for starch food sample place on dropping tile — add a few drops of *iodine* in *KI solution*	−	starch absent: yellow *starch* present: black
Testing for fat food sample extract with alcohol — (a) transfer clear liquid to clean test tube (b) add 1 cm depth of *distilled water*	−	fat absent: clear fat present: cloudy
Testing for protein (Biuret test) food sample extract with water — (a) add 1 cm depth *sodium hydroxide (NaOH)* (b) add a few drops of *copper sulphate ($CuSo_4$)* (*wear safety glasses*)	−	protein absent: blue protein present: purple
Testing for vitamin C food sample extract with water — (a) place in 2 cm^3 syringe (b) add drop by drop to 1 cm depth of *DCPIP* or *PIDCP* (*blue dyes*)	−	Vit. C absent: blue Vit. C present: colourless

THE INDIVIDUAL ORGANISM
3.5 Plants and food: photosynthesis

Green plants are *autotrophs*. *Photosynthesis* occurs mostly in leaves. Leaves are *well adapted* to their function (see page 48).

A dicotyledonous leaf

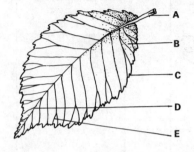

- **A** petiole (stalk)
- **B** leaf blade
- **C** lamina
- **D** midrib (main vein)
- **E** veins

Photosynthesis can be described by the following simplified equations:

(a) Word equation: Carbon dioxide + water $\xrightarrow[\text{sunlight energy}]{\text{chlorophyll}}$ glucose + oxygen

(b) Chemical equation: $6CO_2 + 6H_2O \longrightarrow C_6H_{12}O_6 + 6O_2$

Photosynthesis takes place in the *palisade layer* of leaves. Raw materials, carbon dioxide and water must arrive here. CO_2 arrives via the *stomata* and spongy mesophyll by *diffusion* (some is released during respiration). Water arrives in the xylem of veins from the roots where it was absorbed from the soil by osmosis. *Transpiration, capillarity* and *root pressure* contribute to its movement up the plant. Sunlight energy is used to combine raw materials.

The fate of photosynthetic products
Glucose is quickly converted to *starch* for temporary storage in leaves. At night, starch is broken down again and transported in phloem to other parts of the plant where it is converted and used as:

(a) *Sugars*, as the substrate for respiration (e.g. all cells).
(b) *Starch*, as a temporary storage compound (e.g. carrot root).
(c) *Cellulose*, in growing areas (e.g. cell walls).
(d) *Fats*, as a long-term storage compound (e.g. seeds).
(e) *Proteins* (by addition of nitrates) as enzymes or part of membranes.
(f) *Vitamins* to help enzyme actions.

Oxygen is used in respiration or excreted through stomata.

Stomata

Looking at stomata

A leaf can be plunged into boiling water. The air in the leaf expands and escapes through stomata as air bubbles.

The lower surface of a leaf can be examined under the microscope. A thin layer of nail varnish is painted on to both surfaces, allowed to dry and then peeled off using forceps.

Stomata remain open to allow carbon dioxide in and oxygen out of the leaf during photosynthesis, i.e. in the light. The cells inside the leaf are covered with a layer of water, allowing *diffusion* in and out of them. Water evaporates all the time that stomata are open (*transpiration*). Two *guard cells* are responsible for opening and closing each stoma. Unlike other epidermal cells, they contain chloroplasts and it is thought that their photosynthetic products raise the *osmotic pressure* (O.P.) of cell sap, in the light, opening the stomata. However, the process is very rapid, and other biologists imply that pH is the critical factor, while yet more suggest that *active transport of ions* is involved.

Stoma open
surface view   section

Stoma closed
surface view   section

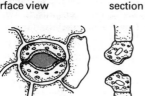
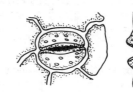

High light intensity;
O.P. of guard cell sap high;
guard cells turgid;
high pH (CO_2 used up).

Low light intensity;
O.P. of guard cell sap low;
guard cells flaccid;
low pH (CO_2 accumulated).

To demonstrate the products of photosynthesis

Glucose can be detected using Benedict's solution (see page 66).

Starch can be detected using the following method:

(a) Plunge leaf into beaker of boiling water for two minutes, to kill leaf and break open cells.
(b) Turn off bunsen, to prevent fire.
(c) Place leaf in boiling alcohol at 80°C, to remove chlorophyll.
(d) Rinse with cold water, to soften leaf.
(e) Place on a white tile and test using iodine solution, to see if starch is present.

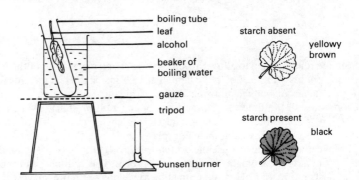

Oxygen can be detected using the following method:

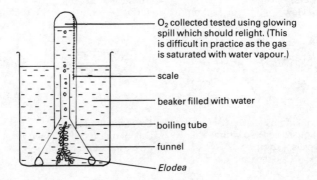

Plants and food: photosynthesis

To demonstrate factors needed for photosynthesis

All demonstrations use destarched plants, i.e. those which have been kept in the dark for at least twelve hours. Iodine solution can then be used to test for the appearance of starch as evidence that photosynthesis has taken place (see page 66).

Carbon dioxide (CO_2)

Test for Starch

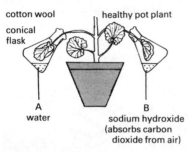

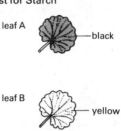

Apparatus left in light for 12 hours

Starch present only where CO_2 available (A)

Sunlight
Masked leaf left in light for 12 hours

Chlorophyll
Variegated leaf left in light for 12 hours

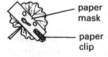

Test for starch

Test for starch

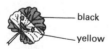

Starch present only in lit areas.

Starch present only where chlorophyll exists.

Plants and food: photosynthesis

Measuring the rate of photosynthesis
The apparatus for collecting oxygen can be adapted to give a measure of the *rate of photosynthesis*. The length of the bubble of air trapped in a given time is recorded under different conditions.

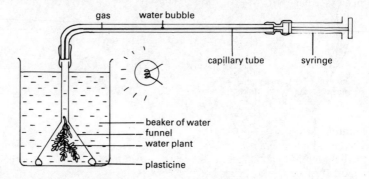

The effects of light intensity can be investigated by placing a light bulb at different distances from the beaker.

The effects of light wavelength (colour) can be investigated by placing coloured filters between the light source and the beaker.

The effects of carbon dioxide concentration can be investigated by filling the beaker with known concentrations of sodium hydrogen carbonate solution.

Photosynthesis and respiration
In all experiments similar to this one, it must be remembered that plant tissue *respires* all the time. This means that the effects recorded are the *net result of photosynthesis and respiration*.

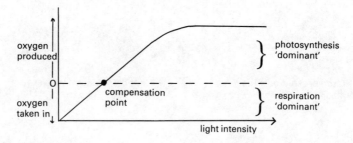

At the *compensation point*, the amount of oxygen produced by photosynthesis equals the amount of oxygen needed for respiration.

Plants and food: photosynthesis 72

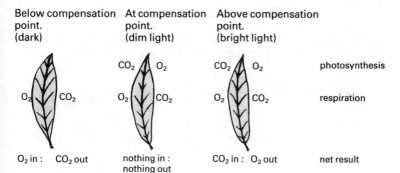

Limiting factors in photosynthesis

The last graph also shows that the curve reaches a point where increasing the light intensity no longer increases the rate of photosynthesis. Another factor must be in short supply and therefore be controlling the rate of reactions. This is called the *limiting factor*. Increasing the level of this factor can raise the rate of photosynthesis. For example, the carbon dioxide level may not be optimum and raising the concentration will then have the effect shown below.

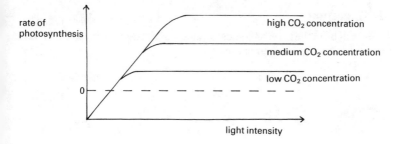

THE INDIVIDUAL ORGANISM
3.6 Animals and food

Animals are *heterotrophs*. They obtain ready-made food which is needed *inside their body cells*. Four stages are involved in getting it there:

1. **Ingestion:** the intake of complex organic molecules into a cavity in the body.

2. **Digestion:** the breakdown of complex organic molecules into simple soluble molecules. This is aided by *secretion* of useful chemicals, e.g. enzymes.

3. **Absorption:** the uptake of soluble food molecules into the body.

4. **Assimilation:** the use of soluble food molecules by body cells. This is followed by

   Egestion: the removal of undigested food from the body.

(In parasites and saprophytes these stages may be modified.)

1 Ingestion

Food can be in the form of liquids, small solids or large solids. Animals have many aids to acquire their chosen food:

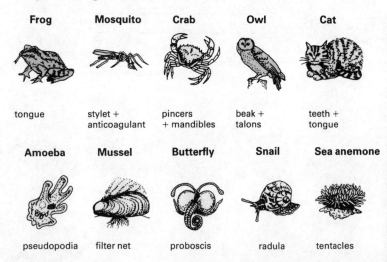

Frog	Mosquito	Crab	Owl	Cat
tongue	stylet + anticoagulant	pincers + mandibles	beak + talons	teeth + tongue

Amoeba	Mussel	Butterfly	Snail	Sea anemone
pseudopodia	filter net	proboscis	radula	tentacles

Animals and food 74

2 Digestion

In mammals, the *alimentary canal*, or *gut*, is responsible for digestion and its structure is modified according to the animal's diet. The *carnivore* gut and *omnivore* gut are relatively *short*, as they deal with easily digestible proteins. The *herbivore* gut is relatively *long*, as it deals with cellulose, which is more difficult to digest.

Alimentary canal labels (for human, rabbit and cow, illustrated below)

A salivary glands
B tongue
C teeth
D mouth
E epiglottis
F trachea
G oesophagus
H stomach
I rumen
J muscle sphincters
K liver
L gall bladder

M bile duct
N pancreas
O pancreatic duct
P duodenum
Q ileum
R small intestine
S caecum
T appendix
U colon (large intestine)
V rectum
W anus

Human: alimentary canal and associated structures

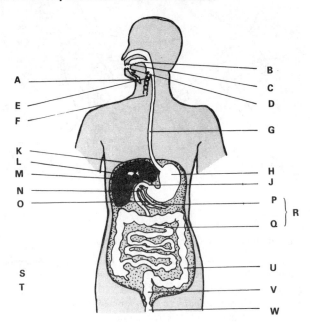

Rabbit: alimentary canal and associated structures

Cow: alimentary canal and associated structures

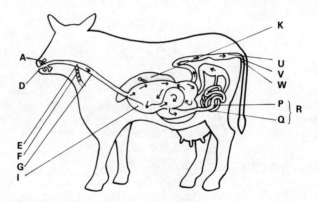

Note that in herbivores, digestion may be helped by:

(a) *An enlarged caecum* (e.g. in rabbit). This contains bacteria which secrete *cellulase* enzymes. (Digested cellulose is passed out with the faeces, which are eaten and absorbed.)
(b) *A four-chambered stomach* (e.g. in cow). One chamber, the *rumen*, contains bacteria which ferment cellulose.

There are *two* types of processes involved in digestion: *physical* and *chemical*.

Animals and food 76

Physical processes of digestion

Chewing and swallowing

A mammal chews its food with its *teeth* and forms it into a soft ball or *bolus*. This is lubricated by *mucus* in *saliva*. It is then swallowed. The floor of the mouth is raised and the *epiglottis* flaps over the trachea. This stops food 'going down the wrong way'. After this, the passage of food through the gut is *involuntary*.

Teeth

Plan of adult human mouth   Key to tooth types

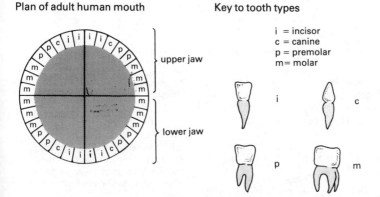

Dental formulae are written as X^*_*: X represents the *type* (incisor, canine, premolar or molar); * represents the number of teeth of this type on the *upper* jaw; $_*$ represents the number of teeth of this sort on the *lower* jaw. The formula is always given for *one side only*; so for the human mouth above the formula is: $i_2^2, c_1^1, p_2^2, m_3^3$

(The four back molars that erupt last to complete an adult set of 32 *permanent* teeth are often called *wisdom teeth*. This set replaces the 20 *deciduous* or *milk teeth* of children that are lost as the jaw grows.)

Animals and food

Tooth structure
Incisor tooth
Molar tooth

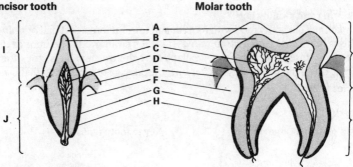

- **A** enamel (hardest substance in body)
- **B** dentine (bone-like)
- **C** nerves in pulp cavity
- **D** blood vessels in pulp cavity
- **E** gum
- **F** jaw bone
- **G** cementum
- **H** periodontal fibres
- **I** crown
- **J** root

Herbivore skull (e.g. sheep)

dental formula:
$$i, c, p, m = \begin{matrix} 0 & 0 & 3 & 3 \\ 3 & 1 & 3 & 3 \end{matrix}$$

The incisors and canines are used with the pad of gum to grip and crop grass. The premolars and molars are ridged to grind grass. The lower jaw moves from side to side.

Teeth never stop growing to replace worn enamel and dentine

Carnivore skull (e.g. dog)

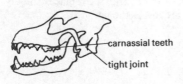

dental formula:
$$i, c, p, m = \begin{matrix} 3 & 1 & 4 & 2 \\ 3 & 1 & 4 & 3 \end{matrix}$$

The incisors are sharp, to bite meat. The canines are long and pointed to kill prey and tear flesh. The powerful carnassials crack bones and act like shears to strip off meat. The lower jaw moves up and down.

Animals and food 78

Tooth decay (dental caries) Bacteria in the mouth feed on *sugar* left on the teeth. These form a mixture called *plaque*. Plaque results in *acid* which *dissolves tooth enamel*. If left undisturbed, plaque hardens to form *tartar*. It also causes gum diseases. Tooth decay takes place in the following stages:

(a) Acid from plaque penetrates enamel	(b) Cavity enlarges	(c) Once in the dentine, decay accelerates	(d) If decay is unchecked, abscess forms

Dentistry If decay is detected at stages (*a*), (*b*) or (*c*), a dentist can remove the decayed part and surrounding tissue and use *amalgam* to make a *filling*. A chipped or decayed crown can be *capped* with metal or *crowned* with porcelain. Where teeth have to be *extracted*, they can be replaced by *dentures*. Tartar can be *scraped* off teeth, and *fissure sealing* (where the tooth surface is coated with a resistant plastic) stops food settling on teeth.

Prevention of decay (looking after teeth and gums)

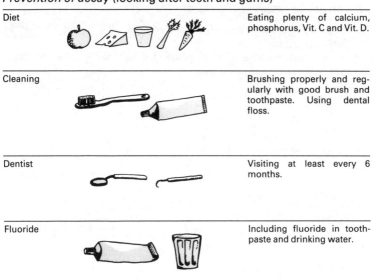

Diet	Eating plenty of calcium, phosphorus, Vit. C and Vit. D.
Cleaning	Brushing properly and regularly with good brush and toothpaste. Using dental floss.
Dentist	Visiting at least every 6 months.
Fluoride	Including fluoride in toothpaste and drinking water.

Peristalsis, churning and constrictions

(a) *Peristalsis* (throughout gut); the gut wall includes a layer of *longitudinal muscle* and one of *circular muscle*, which work *antagonistically* to push food in one direction.

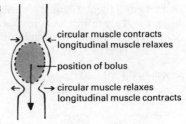

(b) *Churning* (in stomach)

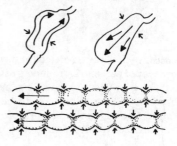

(c) *Constrictions* (in intestines)

Chemical processes of digestion

This is largely effected by *digestive enzymes* which all show similar characteristics (see page 62).

It is important to revise the table below, before looking at the summary of human digestion on the next page.

Enzyme type	Examples	Substrate	Products
Carbohydrases	Sucrase, Lactase, Maltase, Amylase	Disaccharides, Polysaccharides	Monosaccharides (via Disaccharides)
Lipases	Steapsin	Fats	Fatty acids, Glycerol
Proteases	Pepsin, Trypsin, Erepsin	Proteins	Amino acids (via Peptides and Peptones)

3 Absorption

This is often aided by a huge increase in surface area of the gut by:

(a) increase in its length;
(b) presence of *villi* (singular *villus*).

Animals and food

Summary of human digestion and absorption

Mouth — Salivary glands: secrete saliva (slightly alkaline pH) containing salivary amylase

Oesophagus — no digestion takes place here

Stomach — Stomach walls: secrete gastric juice (acid pH) containing HCl, pepsin, and rennin which clots milk in young mammals

Duodenum — Gall bladder: secretes bile (alkaline pH) containing $NaHCO_3$, bile salts which emulsify fats, and bile pigments which colour food brown

Pancreas: secretes pancreatic juice (alkaline pH) containing amylase, steapsin or lipase, and trypsin.

Ileum — Crypts of Lieberkuhn: secrete intestinal juice (neutral pH) containing sucrase, maltase, lactase, and erepsin

- epithelium
- capillary network (carries away water-soluble substances)
- lacteal (carries away fat-soluble substances)
- goblet cells (secrete mucus)
- Crypt of Lieberkuhn

Caecum + appendix — no digestion takes place here
*(In herbivores bacterial cellulase digests cellulose)

Colon — no digestion takes place here. Water is absorbed.

Rectum — no digestion takes place here. Faeces are formed.

Anus — no digestion takes place here. Faeces are egested.

81 *Animals and food*

4 Assimilation

An outline of the ways in which the body uses the absorbed food molecules is given below.

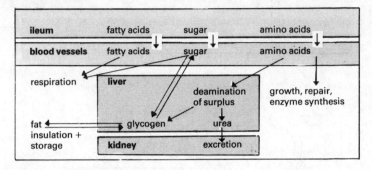

The liver This is the largest organ in the body. It has three *lobes*. It is described by the adjective *hepatic*.

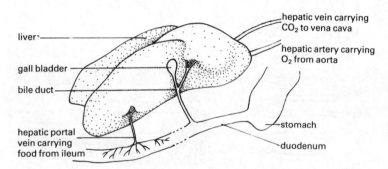

It has many functions and plays a part in:

Level of blood glucose
Iron storage
Vitamin storage
Excess protein deamination, producing *urea* which goes to the *kidneys*
Regulation of body temperature, since it is the main source of body heat
Level of lipids and production of cholesterol
Old red blood cell removal and breakdown
Bile salts production (emulsifying fats in digestion)
Ethanol, other poisons and pathogens being made harmless
Synthesis of blood plasma proteins (e.g. fibrinogen needed for blood clotting).

Making a simple model of digestion

Method

| Fill a length of Visking tubing, knotted at one end, with a fresh starch and saliva mixture | Tie the second end and rinse the tubing with tap water | Place tubing into a boiling tube of distilled water |

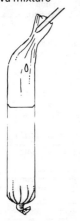  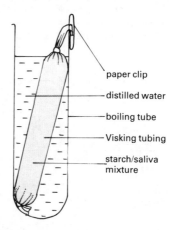

Test the contents of the boiling tube for starch and for sugar (see page 66) at the start of the experiment and again after 20 minutes.

Likely results are shown in the table below:

	Colour with Benedict's	Sugar present?	Colour with Iodine	Starch present?
At start	blue	×	yellow	×
After 20 minutes	orange	√	yellow	×

Explanation The saliva contains an enzyme which converts starch to smaller sugar molecules. This demonstrates *digestion*. The Visking tubing is a *semi-permeable membrane*: it allows small molecules like sugar to pass through it while retaining large molecules like starch. The passage of the sugar products across the Visking tubing demonstrates *absorption*, as in the ileum.

THE INDIVIDUAL ORGANISM
3.7 Transport in flowering plants

There are two transporting fluids in flowering plants, *xylem* and *phloem*. These are contained within *xylem vessels* (*tracheids*) and *phloem sieve tubes* respectively.

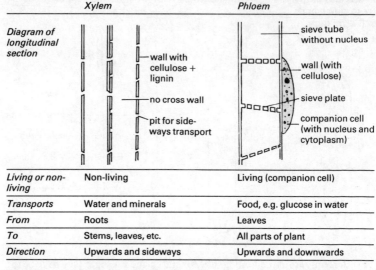

	Xylem	Phloem
Living or non-living	Non-living	Living (companion cell)
Transports	Water and minerals	Food, e.g. glucose in water
From	Roots	Leaves
To	Stems, leaves, etc.	All parts of plant
Direction	Upwards and sideways	Upwards and downwards

The xylem vessels and phloem sieve tubes run side by side. Together with *cambium*, they form *vascular bundles*. The *lignin* in xylem walls and the thick *cellulose* walls of all vascular tissue plays an important part in the *support* of the plant.

Transverse section of root

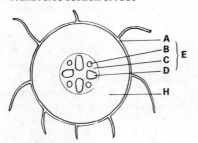

Transverse section of stem

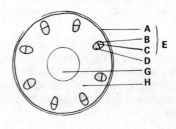

A epidermis
B phloem
C cambium
D xylem

E vascular bundle
F stele
G pith
H cortex

Water uptake, transport and removal in plants

Water enters plant roots from the dilute soil solution by *osmosis*. It is transported up the plant by:

(a) *Transpiration,* largely generated by water evaporating from stomata. This is the most important process.
(b) *Capillarity:* this is attraction between liquid molecules and the sides of very fine xylem vessels which helps water to rise.
(c) *Root pressure,* clearly seen in spring; sap rises up stems before leaves appear and start transpiring.

(Active transport is thought to play a part too.)

Rate of transpiration   This describes how quickly water leaves the leaves. It can be investigated using one of the many types of *potometer*.

Simple mass potometer

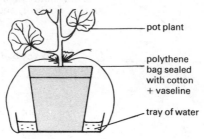

Bubble potometer

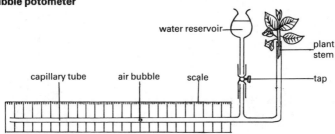

Experiments with potometers show that rate of transpiration *increases* with:

(a) increasing temperature;
(b) decreasing humidity, i.e. drier air;
(c) increasing air movement;
(d) increasing number of stomates open i.e. lighter conditions.

Transport in flowering plants

Summary of water transport through plant

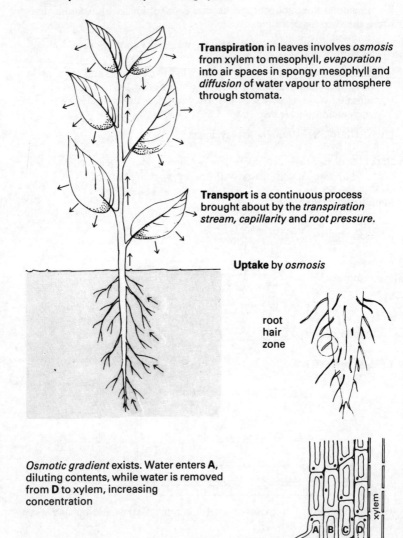

Transpiration in leaves involves *osmosis* from xylem to mesophyll, *evaporation* into air spaces in spongy mesophyll and *diffusion* of water vapour to atmosphere through stomata.

Transport is a continuous process brought about by the *transpiration stream, capillarity* and *root pressure*.

Uptake by *osmosis*

root hair zone

Osmotic gradient exists. Water enters **A**, diluting contents, while water is removed from **D** to xylem, increasing concentration

Food transport in plants

The transport of organic food substances from production sites to the parts where they are needed (*meristems* and storage organs) is called *translocation*. It is not fully understood, but it is thought that living cells are involved since it is affected by:

(a) temperature,
(b) oxygen availability, and
(c) poisons.

To demonstrate that food is transported in phloem

This employs radioactive carbon dioxide ($^{14}CO_2$) which a plant can make into carbohydrates ($[^{14}C_6H_{12}O_6]_n$). This will cause a shadow if rested on a piece of photographic film.

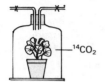

Plant exposed to $^{14}CO_2$ for 24 hours

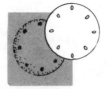

A thin section of stem is rested on photographic film for 10 minutes

Film is compared to stem section preparation under a microscope

It is found that affected areas correspond to phloem.

'Ringing' experiments   Phloem is situated on the inside of the bark of woody plants and is stripped away with it.

At start

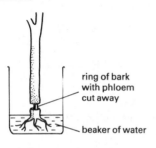

ring of bark with phloem cut away

beaker of water

Several weeks later

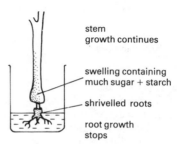

stem growth continues

swelling containing much sugar + starch

shrivelled roots

root growth stops

This experiment would result in the death of the cutting, as food could not reach the roots. These would die and the water supply to the stem would cease.

THE INDIVIDUAL ORGANISM
3.8 Transport in mammals

There are two transporting fluids in mammals: *blood* and *lymph*. These are contained within *blood vessels* (arteries, veins and capillaries) and *lymph vessels* respectively. Blood is pumped by the *heart*.

Studying blood in the laboratory

Centrifugation
Blood is placed in special tubes and spun at high speed in a centrifuge. The cells fall to the bottom.

Microscopic examination
A drop of blood is taken from the thumb using sterile technique and a lancet. A smear is made and stained with Leishman's or Wright's stain. It is viewed under a high power microscope.

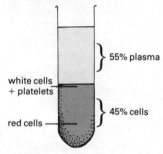

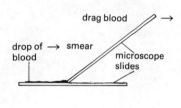

Components of blood and related substances

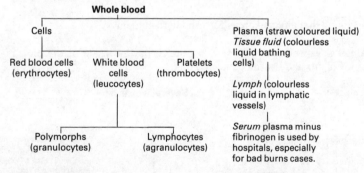

Functions of blood
(a) Transport of food, gases, hormones and urea.
(b) Defence by clotting, engulfing pathogens and antibody production
(c) Heat distribution and temperature control
(d) Tissue fluid formation

Transport in mammals 88

Component	Size	Numbers	Where made	Function
Plasma	—	—	blood vessels	90% water containing dissolved organic foods, minerals, blood proteins, hormones, urea, clotting factors and carbon dioxide (converted to hydrogen carbonate ion HCO_3^-, by carbonic anhydrase for transport).
Red blood cell (no nucleus)	7.5 μm	5,000,000 mm^{-3}	red bone marrow	Red pigment in cells (HAEMOGLOBIN) combines with oxygen at the lungs. Forms oxyhaemogloblin which is taken to respiring cells. Some carbon dioxide carried as carbamine haemoglobin.
Platelets (no nucleus)	2 μm	250,000 mm^{-3}	red bone marrow	Cell fragments that help in blood clotting by releasing THROMBOKINASE prothrombin ⟶ thrombin; fibrinogen ⟶ fibrin; traps cells ⟵ mesh of fibres.
Polymorph lobed nucleus	10 μm	7,000 mm^{-3}	red bone marrow	Engulf invading pathogens. Enzymes digest pathogen. Once full, cells die and often form PUS at the site of infection.
Lymphocyte large nucleus	10 μm	2,000 mm^{-3}	lymph nodes	When pathogens enter cells and multiply, they produce TOXINS which give SYMPTOMS of illness. Lymphocytes make ANTIBODIES specific to pathogens which render them harmless. MEMORY CELLS are left to ward off a second attack by the pathogen.

Blood vessels

The three types of blood vessels – **arteries**, **veins** and **capillaries** – are compared below.

Arteries	Veins
Carry blood *away* from the heart	Carry blood *towards* the heart
Carry *oxygenated* blood (exception = pulmonary artery)	Carry *deoxygenated* blood (exception = pulmonary vein)
Carry blood at *high pressure*	Carry blood at *low pressure*
TS — elastic layer, muscle layer, endothelium	TS — elastic layer, muscle layer, endothelium
LS	LS — valve
Narrow bore (widening at each heartbeat to give *pulse*)	Wide bore
Thick-walled	Thin-walled
No valves present	Valves present
Branch into *arterioles*	Branch into *venules*

Capillaries	
TS, LS — wall one cell thick, endothelium	Thin walls allow *diffusion* in and out of bloodstream. In addition, polymorphs escape and tissue fluid leaks to fight infection in the tissues. Very narrow bore slows blood flow between arteries and veins.

Transport in mammals 90

Naming blood vessels Most arteries and veins are named according to the organ that they serve:

Pulmonary relates to the lungs,
Hepatic relates to the liver,
Renal relates to the kidneys,
Mesenteric relates to the gut.

Check that you can label **A–J** on the diagram of human circulation below.

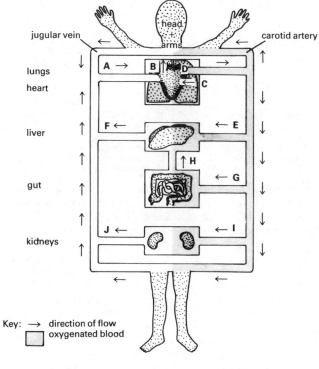

A vena cava	**F** hepatic vein
B pulmonary artery	**G** mesenteric artery
C pulmonary vein	**H** hepatic portal vein
D aorta	**I** renal artery
E hepatic artery	**J** renal vein

Transport in mammals

The heart

Heart seen from the front

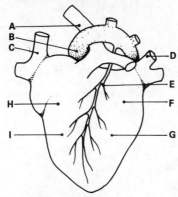

Heart cut open

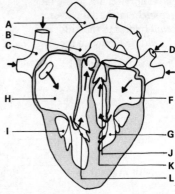

- **A** pulmonary artery (to lungs)
- **B** aorta (to body)
- **C** vena cava (from body)
- **D** pulmonary vein (from lungs)
- **E** coronary artery
- **F** left atrium
- **G** left ventricle
- **H** right atrium
- **I** right ventricle
- **J** semi-lunar valves
- **K** bicuspid valve
- **L** tricuspid valve

(Note that ventricle walls are thicker than atrium walls, as they do the pumping. The left is stronger and thicker than the right to generate the force required to push blood round the entire body.)

The heart is situated in the thoracic cavity between the lungs. Its walls are made of *cardiac muscle*, which *does not fatigue*. The *coronary artery* provides it with food and oxygen (a blood clot in this artery leads to *coronary thrombosis* or heart attack). The heart is protected by the *pericardium* membrane and *rib cage*. It has four chambers: two upper *atria* and two lower *ventricles*. (Remember: blood **a**rrives in the **a**tria and then **trickles** down into the ven**tricles**.)

The heart is best considered as two pumps joined together and working simultaneously. The *right* side receives *deoxygenated blood* from the *body* and pumps it to the lungs for oxygenation. The *left* side receives *oxygenated* blood from the *lungs* and pumps it round the body. As blood travels twice through the heart on a circuit of the body, this is called *double circulation*. *Valves* keep blood flowing in one direction. As they close they make a 'lub-dub' sound that can be heard through a stethoscope.

Average adult heartbeat (pulse) is 60–80 beats per minute.

Average infant heartbeat (pulse) is 100–120 beats per minute.

Pulse *increases* with *exercise* and *adrenalin release* – under strenuous exercise, the heart would be working at a considerably higher rate than 80 beats per minute.

Transport in mammals 92

Blood composition Blood gains and loses substances as it *circulates*:

Gains	Losses
oxygen at lungs	salts, water and urea at kidneys
dissolved food at ileum	carbon dioxide at lungs
urea at liver	oxygen at tissues
carbon dioxide at tissues	dissolved food at tissues
hormones at endocrine gland	glucose at liver
	(stored as glycogen)

From this it is possible to deduce where blood contents are in highest or lowest concentration. For example:

The highest *glucose* concentration is in the *hepatic portal vein*.
The highest *urea* concentration is in the *hepatic vein*.

Lymphatic system

This is a system of tiny blind-ending tubes, called *lymph vessels* or *lymphatics*. These converge on the *thoracic duct*, which empties into the main vein of the left arm. At intervals along the lymph vessels are clumps of cells called *lymph nodes*. These contain stationary *polymorphs* that engulf passing bacteria and tissue producing *lymphocytes* and *antibodies*. Lymph nodes in the neck, armpits and groin are grouped to form *lymph glands*. These swell when the body is infected.

The four functions of the lymphatic system are:
(a) to produce lymphocytes and antibodies;
(b) to help to return excess tissue fluid to blood;
(c) to absorb fats;
(d) to filter out bacteria in lymph nodes.

Blood, tissue fluid and lymph

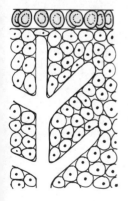

Vessels	Contents	Composition
Capillary (heart + blood vessels)	Blood (filtered)	Contains cells, proteins and fat droplets
Cells with tissue fluid (No cell is more than 25µm away from a capillary)	Tissue fluid (collected and additions made)	No cells, proteins or fat droplets; contains glucose and O_2
Lymph vessel (+ nodes)	Lymph	No cells; contains fat droplets and some proteins but less glucose and O_2

THE INDIVIDUAL ORGANISM
3.9 Respiration

What is respiration?

Respiration is a characteristic of *all living plant and animal cells*. It is a whole series of chemical reactions controlled by *enzymes*. *Energy* is released from organic molecules, usually *glucose*. This energy is stored in energy molecules called ATP (adenosine triphosphate). Each ATP molecule can lose a phosphate group and release the energy that held it on (the bond energy) when it is needed. This leaves ADP (adenosine diphosphate).

$$(A)-(P)-(P)\vdots(P) \xrightleftharpoons[\text{energy needed by organism}]{\text{energy from respiration}} (A)-(P)-(P) + (P)$$

energy-rich bond

Energy **might** be needed for:
Muscle action or other mechanical work;
Impulse along nerves or other electro-chemical work;
Growth or other chemical work;
Heat maintenance, as in warm-blooded animals;
Transporting substances against a concentration gradient.

There are two sorts of respiration, *aerobic* and *anaerobic*, and these are compared below:

	Aerobic	*Anaerobic*
Is oxygen needed?	Yes	No
Where it occurs	Mitochondria of plant and animal cells	Cytoplasm of plant and animal cells
Breakdown	Complete	Incomplete
Amount of energy released from one glucose molecule	A lot (40 ATP molecules)	A little (2 ATP molecules)
Products	CO_2 and H_2O	Lactic acid in animals, alcohol and CO_2 in yeast, other products in bacteria.
Chemical equations	$C_6H_{12}O_6 + 6O_2$ $\rightarrow 6CO_2 + 6H_2O$	$C_6H_{12}O_6 \rightarrow 2C_3H_6O_3$ $C_6H_{12}O_6 \rightarrow 2C_2H_5OH + 2CO_2$.

Some bacteria can only respire anaerobically. Most plants and animals, including yeasts and tapeworms, respire aerobically when oxygen is available and anaerobically when deprived of oxygen.

Examples of anaerobic respiration

Yeast

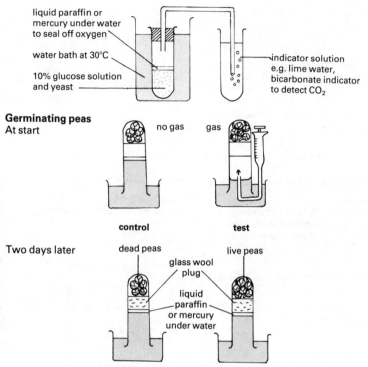

Germinating peas

* A syringe containing indicator solution can be used to test for CO$_2$.

A further example occurs in human muscles which normally respire aerobically. During heavy exercise requiring a lot of energy, oxygen supply to the muscles may not be sufficient and *anaerobic* respiration takes over, producing *lactic acid*. An *oxygen debt* is built up, since oxygen is needed to get rid of the acid. This is 'paid off' by continued aerobic respiration after the exercise has finished.

Oxygen is taken from the atmosphere of terrestrial organisms and usually from the water surrounding aquatic organisms. Notable exceptions are the marine mammals such as whales and diving beetles that cannot use dissolved oxygen, but collect it from the surface.

THE INDIVIDUAL ORGANISM
3.10 Gas exchange

All organisms need energy which is liberated by respiration. Plants, which do not move, need less energy than animals, which move (and may be warm-blooded). Energy requirements influence how living things obtain and release gases for respiration. This is called *gas exchange*.

Features of gas exchange surfaces
Larger plants and animals have special surfaces for gas exchange. They all:

(a) have a large surface area;
(b) are thin to allow gas diffusion;
(c) are moist;
(d) are well ventilated (supplied with gases);
(e) have transport systems to carry gases to and from them.

Simplified diagrams of different gas exchange surfaces.

Plants
Fewer examples exist since much oxygen is supplied as the waste product of photosynthesis.
$$6CO_2 + 6H_2O \rightarrow C_6H_{12}O_6 + 6O_2$$

Filaments (e.g. fungi and algae)

Spongy mesophyll (e.g. flowering plants)

Lenticels (e.g. flowering plants)

Animals
Flattened body (e.g. flatworms)

External gills (e.g. lugworms)

Internal gills (e.g. fishes)

Lungs (e.g. mammals and birds)

Tracheal tubes (e.g. insects)

Specific examples of gas exchange surfaces

Fish (e.g. Cod) Exchange at minutely-branched gill filaments. Water enters the mouth, is forced over the gills and leaves, under pressure, from the operculum.

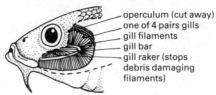

Bird (e.g. Sparrow) Exchange at minute air passages in lungs contracted twice at each breath. Efficiency needed for flight.

Insect (e.g. Honey Bee) Air is supplied directly to cells. Larger insects pump their abdomens to squeeze air in and out of the tracheal system.

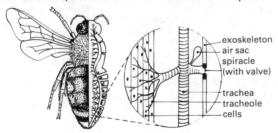

Amphibian (e.g. Common Frog) Exchange at mucus-covered skin. When active, 'gulps' air into lungs. No diaphragm.

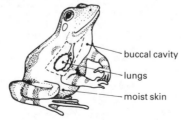

Gaseous exchange

Human gas exchange

Humans are among the animals which breathe, i.e. make movements that bring a source of oxygen to a surface for gas exchange. Models can be used to demonstrate breathing.

Human lungs and associated structures   **Model of human lungs**

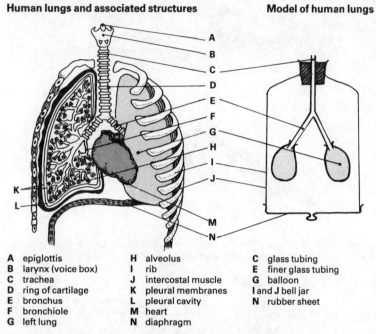

A epiglottis	**H** alveolus	**C** glass tubing
B larynx (voice box)	**I** rib	**E** finer glass tubing
C trachea	**J** intercostal muscle	**G** balloon
D ring of cartilage	**K** pleural membranes	**I** and **J** bell jar
E bronchus	**L** pleural cavity	**N** rubber sheet
F bronchiole	**M** heart	
G left lung	**N** diaphragm	

(The shared letters indicate how parts of the model correspond to structures in the human chest.)

Gaseous exchange 98

Breathing

Inhalation: breathing in
(sometimes called *inspiration*)

(a) diaphragm contracts and moves down
(b) rib cage moves up and out
(c) chest volume increases
(d) chest pressure decreases
(e) air rushes in

Exhalation: breathing out
(sometimes called *expiration*)

(a) diaphragm relaxes and moves up
(b) rib cage moves down and in
(c) chest volume decreases (lungs are elastic)
(d) chest pressure increases
(e) air is pushed out

When revising, you need to learn only inhalation, as exhalation simply reverses each event.

Gas exchange in lungs and tissues

Gas exchange occurs in the 500 million tiny air-sacs or *alveoli* clustered at the end of the bronchioles. There is a strong concentration gradient between alveolar air and the blood in the capillaries. Gas exchange also occurs between body tissue capillaries and cells.

CO_2 is carried as bicarbonate ion (HCO_3^-) in plasma
O_2 is carried as oxyhaemoglobin in red blood cells

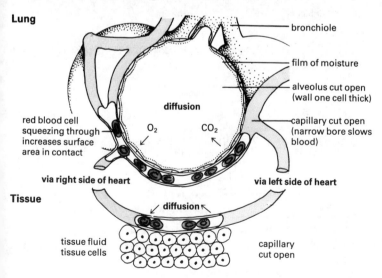

Measuring breathing

There are two parameters (measurements) that are commonly recorded:

(a) *Rate of breathing*, i.e. the number of breaths per minute;
(b) *Depth of breathing*, i.e. the amount of air (in litres) taken in and out at each breath.

A *kymograph* can record these measurements, made using a *spirometer*.

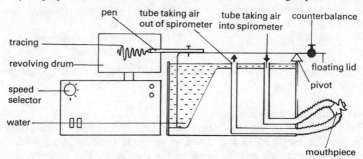

Imagine inhaling through the mouthpiece. The floating lid will sink and the pen will mark a downward stroke on the kymograph drum. If the drum is revolving, one breath (inhaling and exhaling) will look like this:

Examine the tracings below (text books often print such kymograph tracings upside down):

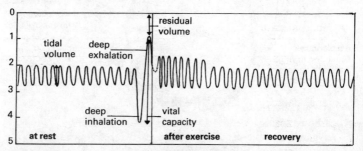

The height of the tracing shows the *depth* of breathing. The number of spikes in a given distance at a known speed setting is used to find the *rate* of breathing.

Typical breathing measurements are:

(a) Rate of breathing: 15–20 breaths per minute
(b) Depth of breathing: 0.5 dm^3

Exercise increases both the rate and depth of breathing.

Breathing rate is controlled by the *respiratory centre* in the *medulla of the brain*. This senses the carbon dioxide level in the blood.

Proportions of the main gases inhaled and exhaled

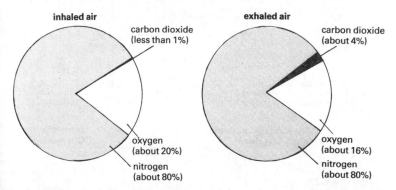

Gases like hydrogen and helium are so rare in the air that special apparatus is needed to detect them. Gases like carbon monoxide, sulphur dioxide and hydrogen sulphide may be present in polluted atmospheres. The amount of water vapour varies in the air we breathe *in*, but it always saturates the air we breathe *out*.

Respiration experiments

Three common experiments are illustrated on page 101. Carbon dioxide production is the most usual means of detecting aerobic respiration.

Suitable indicator solutions and colour changes are:

Indicator name	Without CO_2	With CO_2
Lime water ($Ca(OH)_2$)	Clear	Cloudy/milky
Hydrogen carbonate indicator	Red	Yellow

Other evidence that might be looked for is that:

(a) carbohydrate is used up (using radioactive carbon ^{14}C);
(b) oxygen is used up (using pyrogallic acid to absorb oxygen);
(c) water is evolved (using cobalt chloride);
(d) energy is released (using a thermometer to detect waste heat energy).

101 Gaseous exchange

Experiment 1

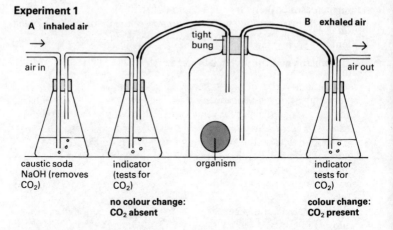

Experiment 2

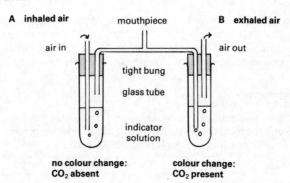

Experiment 3

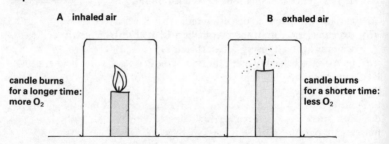

Breathing problems

Asphyxia: lack of oxygen when a person has stopped breathing after an accident (e.g. drowning or electric shock).

Asthma: contraction of the bronchioles caused by an allergic reaction making breathing difficult. The sufferer wheezes.

Bronchitis: inflammation of the bronchial tubes.

Emphysema: breakdown of the delicate alveolar walls.

Hay fever: an allergic reaction to pollen where the body's immune system over-reacts. The nose lining becomes sensitive and inflamed, and produces a lot of mucus so the person gets a runny nose and sneezes.

Laryngitis: inflammation of the voice box causing loss of voice.

Lung cancer: a growth (tumour) develops in the wall of the bronchial tubes.

Pleurisy: inflammation of the pleural membranes which is particularly painful.

Pneumonia: fluid collects in the alveoli due to severe infection of the lungs.

Tuberculosis (TB): a bacterial infection which destroys lung tissue. TB, or consumption, as it was called, used to be a major killer.

Smoking Smoking is thought to be an important cause of:
(*a*) lung and other cancers; (*b*) chronic bronchitis; (*c*) emphysema; (*d*) heart disease; (*e*) stomach ulcers.

Pregnant women who smoke are more likely to give birth to undersized babies, or to have a miscarriage or stillbirth.

A 'smoking machine'

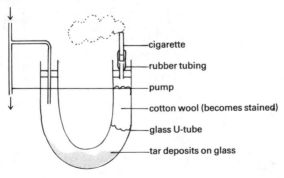

Restoring breathing If brain cells are deprived of oxygen for longer than three minutes they are permanently damaged. *Artificial breathing* (mouth-to-mouth resuscitation or the 'kiss of life') may restore breathing. If this does not succeed a *resuscitator* can keep a patient alive. Where chest muscles are paralysed, an *iron lung* can be used.

THE INDIVIDUAL ORGANISM
3.11 Excretion

Excretion is the removal of the *waste products of metabolism* from an organism. It prevents the build up of substances, often *poisons*, which may interfere with the normal functioning of the body.

Outline of excretion in flowering plants

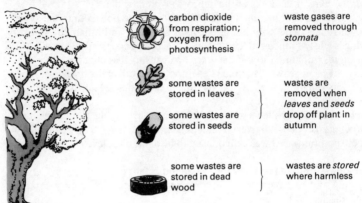

	carbon dioxide from respiration; oxygen from photosynthesis	waste gases are removed through *stomata*
	some wastes are stored in leaves	wastes are removed when *leaves* and *seeds* drop off plant in autumn
	some wastes are stored in seeds	
	some wastes are stored in dead wood	wastes are *stored* where harmless

Outline of excretion in mammals

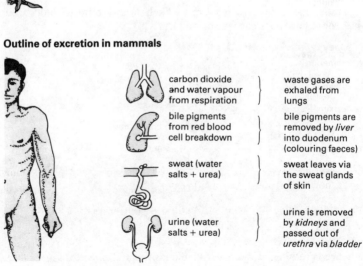

	carbon dioxide and water vapour from respiration	waste gases are exhaled from lungs
	bile pigments from red blood cell breakdown	bile pigments are removed by *liver* into duodenum (colouring faeces)
	sweat (water salts + urea)	sweat leaves via the sweat glands of skin
	urine (water salts + urea)	urine is removed by *kidneys* and passed out of *urethra* via *bladder*

Note: Excretion should not be confused with egestion (the removal of indigestible matter that passes through the gut and out as faeces).

Excretion in humans

How urine is formed Proteins which are eaten but not required cannot be stored. They are *deaminated* in the *liver*. Here, the *ammonia* formed combines with *carbon dioxide* to make less poisonous *urea*. This is carried by the blood to the *kidneys* for excretion.

Human excretory system

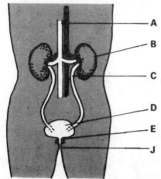

Vertical section through left kidney

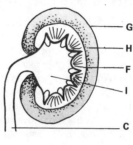

A renal artery and vein
B left kidney
C ureter
D bladder
E sphincter (ring of muscle)
F pyramid of medulla
G capsule
H cortex
I pelvis
J urethra

A kidney tubule (nephron)

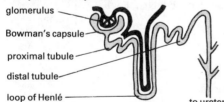

Composition of urine 96% water; 2% urea; 2% salts; some bile pigments; drugs, hormones, poisons; (sugar in cases of diabetes).

There are between one and two million nephrons in each kidney. Each fulfils two roles:

(a) *Ultrafiltration:* all small molecules are filtered out of the blood at high pressure in the *glomerulus* into the *Bowman's capsule*. Blood cells and proteins are held back.
(b) *Reabsorption:* useful small molecules are taken back into the blood from the tubule. This uses up energy and is an example of *active transport*. Harmful substances are not reabsorbed.

THE INDIVIDUAL ORGANISM
3.12 Sensitivity

In order to survive, each living thing detects *stimuli* (changes in the environment) and makes *responses*. This usually means moving all or part of its body to a more suitable environment. This is the basis of *behaviour*.

What plants detect
Plants detect and respond to the following stimuli:

(*a*) *Light:* shoots grow towards light, while roots grow away from it.
(*b*) *Touch:* some shoots grow towards solid objects;
(*c*) *Gravity:* shoots grow away from it, while roots grow towards it;
(*d*) *Water:* shoots grow away from it, while roots grow towards it.

} in the *external* environment

(*e*) *Chemicals:* in the *internal* environment

How plants detect stimuli Little is known of how plants detect stimuli. Shoot tips detect light, but plants have no obvious sense organs.

What animals detect
Animals detect and respond to the following stimuli:

(*a*) *Light* (intensity/brightness + wavelength/colour)
(*b*) *Sound* (volume/loudness + pitch/high or low)
(*c*) *Touch*
(*d*) *Change in temperature*
(*e*) *Pain*
(*f*) *Chemicals*
(*g*) *Gravity* (change in position)

} in the *external* environment

(*h*) *Chemicals*, e.g. CO_2, hormones, sugar
(*i*) *Change in blood temperature*
(*j*) *Change in blood osmotic potential*
(*k*) *Stretching in bones, muscles, ligaments*

} in the *internal* environment

How animals detect stimuli Many animals have highly specialised sense organs: the *eyes, ears, taste buds, receptors of nose lining* and *sensory nerve endings of skin* detect external stimuli; the *hypothalamus, cells in blood vessel walls* and *proprioceptors* detect internal stimuli.

The eye and sight

Section through a human eye

A optic nerve
B blind spot
C retina
D fovea
E choroid layer
F sclerotic
G eye muscle
H eyelid
I eyelash
J conjunctiva
K cornea
L iris
M pupil
N lens
O aqueous humour
P vitreous humour
Q ciliary body/muscle
R suspensory ligament

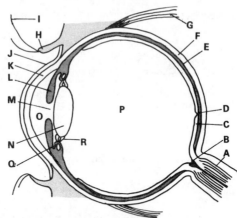

Functions of parts of eye

The *optic nerve* (**A**) carries nerve impulses to the brain.

The *blind spot* (**B**) is the point where optic nerve and blood vessels leave the eye.

The *retina* (**C**) contains light-sensitive cells of two types:
 (*a*) Rods, which detect black and white in dim light;
 (*b*) Cones, which detect colour in bright light.

The *fovea* (**D**) only contains cones, densely packed to give very accurate vision.

The *choroid layer* (**E**) is a black layer preventing reflection and carrying blood vessels transporting food and oxygen to the eye.

The *sclerotic* (**E**) is a tough protective outer layer maintaining the shape of the eye.

The *eye muscle* (**G**) is attached to the sclerotic to move the eyeball.

The *eyelid* (**H**) blinks to protect against very bright light and to move bactericidal tears over the cornea.

The *eyelash* (**I**) (together with the eyebrows) prevents falling dust entering the eyeball.

The *conjunctiva* (**J**) protects the cornea.

The *cornea* (**K**) bends light rays towards the retina.

The *iris* (**L**) is the coloured part of the eye which alters the size of the pupil.

The *pupil* (**M**) is the hole which allows light to enter the eye.

The *lens* (**N**) changes shape to focus an image on the retina.

The *aqueous humour* (**O**) is watery and supports the lens.

The *vitreous humour* (**P**) is jelly-like and helps the sclerotic to maintain the eyeball's shape.

The *ciliary body/muscle* (**Q**) alters lens shape during accommodation.

The *suspensory ligament* (**R**) attaches the ciliary muscles to the lens

Controlling the amount of light entering the eye

In dim light

Radial muscles of iris *contract*
Circular muscles *relax*
Pupil becomes larger
More light enters

In bright light

Radial muscles of iris *relax*
Circular muscles *contract*
Pupil becomes smaller
Less light enters

Focussing light (accommodation)

The process of focussing

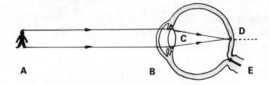

| **A** Light rays bounce off an object | **B** Rays are bent passing through cornea | **C** Rays are bent passing through lens | **D** Focused image is formed upside down on retina | **E** optic nerve carries message to brain |

Seeing near objects	*Seeing distant objects*
Ciliary muscle contracts Suspensory ligaments slacken Lens returns to normal thick shape Pupil is small	Ciliary muscle relaxes Suspensory ligaments tauten Lens is stretched thin Pupil is large

Vision in other animals

Humans and some other primates are the only mammals with colour vision. Cats, dogs and even bulls cannot distinguish colours. Birds that hunt by day and certain insects like bees also see 'in colour'. Prey animals, such as rabbits, have eyes at the sides of their heads. They can see through 360° without moving. Predatory and tree-dwelling animals have eyes at the front of their heads. This allows *judgement of distance*, and is called *stereoscopic* (binocular) vision.

Sensitivity 108

Problems with vision

Short-sightedness (myopia)

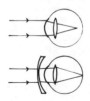

(a) focus is in front of the retina;
(b) eyeball is too long;
(c) distant objects are blurred e.g. TV;
(d) corrected by concave (diverging) lens in spectacles or contact lenses.

Long-sightedness (hypermetropia)

(a) focus is behind the retina;
(b) eyeball is too short or cornea and lens too 'weak';
(c) near objects are blurred e.g. books;
(d) corrected by convex (converging) lens in spectacles or contact lenses.

Old-sightedness (presbyopia): long-sightedness; lens becomes inflexible.
Cataracts: cloudiness of lens
Astigmatism: cornea or lens mis-shapen
Colour blindness: red and green are confused; this is found in 1 in 12 males and in 1 in 200 females.

The skin and 'feeling'

Skin structure is illustrated on page 49. Different sensory cells detect the following stimuli:

pain

pressure

touch

temperature change

Some parts of the body have more touch-sensitive sensory cells than others. Finger tips, earlobes and lips are most sensitive. Soles of feet, elbows and knees are least sensitive. This can be shown as follows:

hairpin ends 5 mm apart

Blindfolded subject is gently touched by a hairpin on different parts of body. If nerve endings are more than 5 mm apart, only one end is felt.

The sense of touch is very important, especially to blind people whose fingertips replace eyes in 'reading' *Braille*.

Human ear and hearing

The ear is divided into three sections, as shown below.

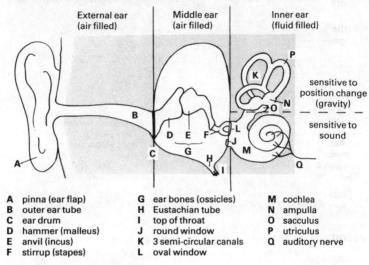

A pinna (ear flap)	G ear bones (ossicles)	M cochlea
B outer ear tube	H Eustachian tube	N ampulla
C ear drum	I top of throat	O sacculus
D hammer (malleus)	J round window	P utriculus
E anvil (incus)	K 3 semi-circular canals	Q auditory nerve
F stirrup (stapes)	L oval window	

The ear detects *sound waves*

Amplitude determines *volume*:
Frequency determines *pitch*:

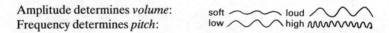

How the ear works

Outer ear collects sound waves

Ear ossicles amplify sound 27 times

Cochlea (uncoiled) converts sound waves to nerve impulses by stimulating sensory nerve endings in the *organ of Corti*

Auditory nerve takes impulses to the brain

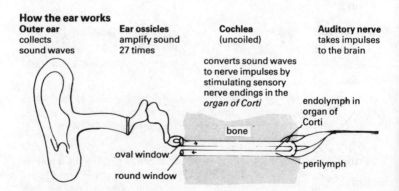

Hearing in other animals The pinna, which collects sound waves, is very large, relative to body size, in some species (e.g. rabbit, elephant). A large pinna may also help to lose heat or get rid of flies.

Sensitivity 110

The ear and balance

Organs of balance are situated in the semi-circular canals of the inner ear. There are three placed at right angles to each other.

Inside the utriculus (detects changes in position)

Inside the ampulla (detects movements)

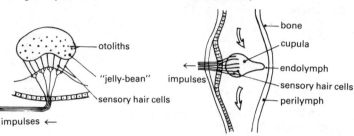

Information from the eyes generally combines in the brain with messages from the semi-circular canals.

The tongue and taste

Our tongues can only detect *four* basic tastes, each with a different type of *taste bud*. They are:

1 Salt   2 Sweet   3 Sour   4 Bitter

and are distributed on the tongue as shown on the map. Sensory cells are stimulated by chemicals dissolved in saliva. The sense probably serves only to tell us if food is suitable to eat.

Map of tongue

Section through a taste bud

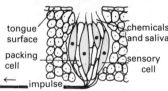

The nose and smell

Most of the 'flavour' of our food comes from our sense of smell. (Think how food loses its flavour when your nose is blocked with a cold.) Chemicals in the air dissolve in the moist lining of the nose. Sensory cells situated here can detect very low concentrations of chemicals and are much more sensitive than any machines yet invented.

THE INDIVIDUAL ORGANISM
3.13 Response and Structure

Plant responses
Plant responses involve different rates of growth (or sometimes more rapid changes in turgidity, e.g. stomata opening and closing and hinge cells). Responses are put into two groups: *tropic* responses, or *tropisms*, and *nastic* responses.

Tropisms
Tropisms are growth movements in which the direction is determined by the stimulus. They are either *positive* where the movement is *towards* the stimulus, or *negative* where the movement is *away from* the stimulus. The following prefixes are also used:

Geo-, where the stimulus is *gravity*
Hydro-, where the stimulus is *water*
Thigmo-, where the stimulus is *touch*
Photo-, where the stimulus is *light*

Investigating regions of response A growing shoot is marked with equally-spaced lines and left in unidirectional light. The region of response occurs *behind the growing tip* in the region of elongation. Cells on one side have grown more than on the other due to uptake of water and *vacuolation* (formation of vacuoles).

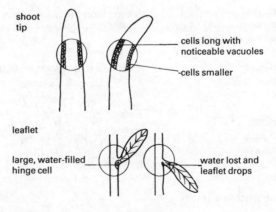

Nastic responses
Plants also respond to a variety of stimuli that do not come from any particular direction, such as general lighting, temperature and humidity. The opening and closing of flowers is an example.

Animal responses

Responses of higher animals usually involve *movement*. Fewer are in the form of *secretions from glands* (e.g. saliva production by human salivary glands when food is placed in the mouth). Simple invertebrates, protozoans and even individual animal cells (e.g. white blood cells and sperms) are said to demonstrate *taxis* or *kinesis*.

Taxis is movement *towards* or *away from* a stimulus (e.g. earthworms move towards formalin or detergent in soil).

Kinesis is movement *at random* in response to a stimulus (e.g. woodlice in dry or light areas run around and only stop when in damp, dark areas).

Skeletons

The *support* which animals require to generate movement comes from *skeletons*. Skeletons resist compression and provide a rigid framework against which *muscles* can act. They often offer *protection* also. There are at least three types:

Hydrostatic skeleton

'Soft-bodied' animals are supported mainly by *liquid* in their bodies. Muscles contract against the liquid, changing its shape and thus that of the body. The pressure of the liquid against muscles gives some rigidity, e.g. earthworm, slug, caterpillar.

Exoskeleton

Some animals are supported by hard, dead material which forms in plates on the *outside* of their bodies. Plates are connected by flexible *membranes* at *joints* and have *inward projections* for muscle attachment. During growth, the exoskeleton is shed and replaced at intervals (*ecdysis*) to allow increases in body size, e.g. crab, prawn and woodlouse (skeletons incorporate lime); beetles, bees and ants (skeletons incorporate chitin).

Endoskeleton

Many animals are supported by a hard framework of living *cartilage* or *bone, inside* their bodies. These are attached by *ligaments* at *joints* and have *external ridges* for *muscle attachment*, e.g. mammals, birds, reptiles, amphibians and fishes (skeletons incorporate calcium and phosphorus salts).

Human skeleton

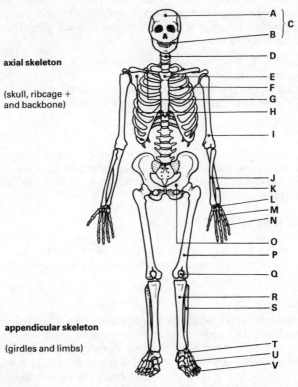

axial skeleton

(skull, ribcage + and backbone)

appendicular skeleton

(girdles and limbs)

- **A** cranium
- **B** mandibile
- **C** skull
- **D** vertebrae
- **E** sternum
- **F** ribs
- **G** clavicle
- **H** scapula
- **I** humerus
- **J** radius
- **K** ulna
- **L** carpals
- **M** metacarpals
- **N** phalanges/digits
- **O** pelvis
- **P** femur
- **Q** patella
- **R** tibia
- **S** fibula
- **T** tarsals
- **U** metatarsals
- **V** phalanges/digits

Bone structure

Section through a long bone

Microscopic structure of bone

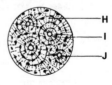

- **A** cartilage
- **B** dense bone
- **C** spongy bone
- **D** red bone marrow
- **E** white bone marrow
- **F** head (epiphysis)
- **G** shaft
- **H** calcium + phosphorus deposits
- **I** bone cell
- **J** Haversian Canal (with blood vessels)

Joints Bones meet at *joints*. Joints are classified according to the amount of movement they allow. Starting with those allowing most movement, the different types of joint are:

(a) *Ball and socket* (movement in all planes and rotational), e.g. hip, shoulder;
(b) *Hinge* (movement in one plane), e.g. knee, elbow;
(c) *Pivot* (rotational movement), e.g. between atlas and axis, between radius and ulna at wrist;
(d) *Gliding* (sliding movement), e.g. between vertebrae, in foot and hand;
(e) *Fused* (no movement), e.g. skull and pelvis.

The most movable joints – ball and socket, and hinge – are sometimes called *synovial joints*.

Typical synovial joint (simplified)

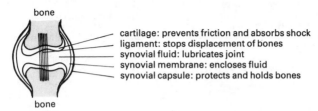

The conditions *housemaid's knee* and *tennis elbow*, which involve fluid build-up in the tissues surrounding joints, both arise from wear on the *synovial membrane* and *capsule*. These rupture and the synovial fluid leaks out.

115 *Response and Structure*

Axial skeleton The *skull* and *rib cage* each have an important role in *protection*. The skull encloses the brain and the organs of sight, smell and hearing. The rib cage encloses the heart and lungs.

The *vertebral column* or *spine* consists of 33 bones called *vertebrae*. Some are *fused* (joined together). It has three functions:

(*a*) to protect the spinal cord;
(*b*) to give the trunk some flexibility (by means of cartilage *discs* between bones);
(*c*) to determine posture.

Vertebral column   **Single vertebra** (simplified)

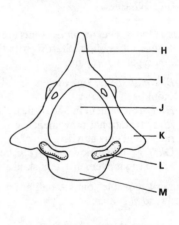

- **A** atlas (allows head nodding)
- **B** axis (allows head shaking)
- **C** 7 cervical
- **D** 12 thoracic (support ribs)
- **E** 5 lumbar (supports trunk muscles)
- **F** 5 sacral or sacrum (fused with pelvis)
- **G** 4 caudal or coccyx (tail of other mammals)

- **H** neural spine (for muscle attachment)
- **I** neural arch (protects cord)
- **J** neural canal (contains cord)
- **K** transverse process (for muscle attachment)
- **L** facet (for articulation)
- **M** centrum (provides strength)

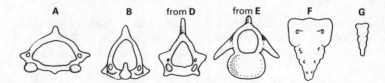

Response and Structure 116

Appendicular skeleton Two *girdles* support the *limbs*:

(a) The *pectoral girdle* (2 scapulas + 2 clavicles) supports *fore-limbs*.
(b) The *pelvic girdle* (pelvic bones + sacrum) supports *hind-limbs*.

Each limb has a typical vertebrate *pentadactyl* (five finger) plan, e.g. in human:

Arm (fore-limb)

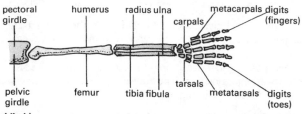

Leg (hind-limb)

Muscle

Bones are moved by contraction of *muscles*. There are three types, but only the first is involved in *voluntary* movement of the skeleton.

Type	Description	Where found	Notes
Striped/skeletal/ voluntary	Striped muscle fibre made up of joined cells	Limbs, jaw, trunk, etc.	Stimulated by nerve impulses from cerebrum (i.e. under conscious control). Contracts powerfully. Soon fatigues.
Cardiac	Striped, separate branching cells	Wall of heart only	*Myogenic* – contracts powerfully without stimulation. Does not fatigue.
Smooth/visceral/ involuntary	Unstriped, separate cells	Walls of gut, blood vessels, bladder, uterus, etc.	Stimulated by nerve impulses from cerebellum (i.e. not under conscious control). Contracts less powerfully. Fatigues slowly.

How striped muscles move bones Striped muscles are firmly attached to bones by tough *inelastic tendons*.

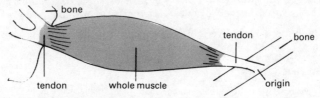

A whole muscle includes thousands of muscle fibres which are able to *contract* (shorten and thicken) or *relax*. Contraction uses *energy* from respiration to exert a *force* on a bone and to move it. Relaxation does *not* reverse this effect. *Gravity* or, more often, *another muscle* is used to return a bone to its original position. Thus muscles usually occur in *antagonistic pairs* (i.e. two muscles exerting opposite forces).

Flexor muscles move bones to *bend* joints.
Extensor muscles move bones to *straighten* joints.

When a flexor contracts, an extensor relaxes and vice versa:

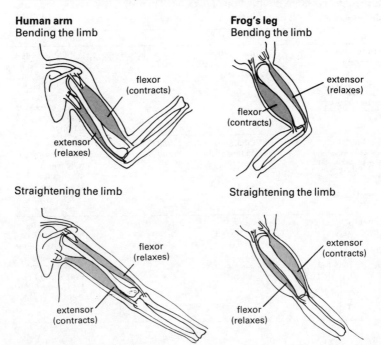

Different types of movement

Swimming (e.g. in fishes)

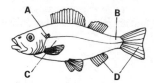

- **A** overlapping scales + mucus (for streamlining)
- **B** muscular tail (muscle blocks called *myotomes* propel fish forward by figure of eight movements)
- **C** internal swim bladder (inflates to give buoyancy; deflates to make fish sink)
- **D** fins give stability from:
  rolling   pitching   yawing

and help steer and brake.

Flying (e.g. in birds)

- **A** hollow bones (for lightness)
- **B** minor pectoral muscle (contracts for upstroke)
- **C** major pectoral muscle (contracts for downstroke). One third body weight.
- **D** keel-like sternum (for muscle attachment)
- **E** feet and legs held up (for streamlining)
- **F** wing bones (extended and fused to give large span)
- **G** overlapping feathers (for large surface area and streamlining)
- **H** tail (works with wings to steer and brake)
- **I** lungs and air sacs (provide very efficient gas exchange)

Birds are heavier than air. Both *gliding* and *flapping flight* need *lift*. Lift is generated by *aerofoil-shaped wings*. When such a shape moves forward, higher pressure is created below the wing, generating lift:

fast air flow (low pressure)

slow air flow (high pressure)

THE INDIVIDUAL ORGANISM
3.14 Co-ordination

Plants and animals are capable of detecting many different stimuli and of making many different responses. *Linkage systems* connect and co-ordinate *receptors* and *effectors*. It is critical that an appropriate response is made to a stimulus.

Co-ordination in plants

Plants secrete *hormones* to co-ordinate their responses. They are made in very small amounts in the roots and shoots, and *diffuse* to the site of action. This partly explains plants' slowness of response. *Auxins* (e.g. IAA) are the most important and, at the site of cell elongation, they control water uptake and *vacuolation* (the quickest means of a cell increasing in size).

Auxin experiments

1 This shows that a hormone (which can diffuse and promote growth) is made in coleoptile tips.

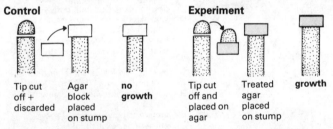

2 This experiment shows how unidirectional light affects hormone distribution and growth.

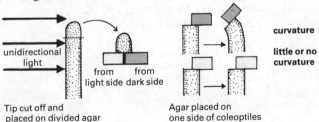

Other hormone effects Hormones control *budding, flowering, fruit-setting, leaf fall* and *seed dormancy*. Horticulturalists use synthetic hormones to increase crop yields and to act as *selective weed-killers* in fields of narrow-leaved cereals or lawns of grass. Broad-leaved weeds absorb more hormones, 'outgrow' themselves and die.

Co-ordination in animals

Higher animals have *nervous systems* and *endocrine* (*hormone*) *systems* to co-ordinate their responses.

Nervous system

The nervous system comprises millions of interconnecting nerve cells called *neurons*. Messages travel very fast along neurons, in the form of *impulses*. The nerves extending throughout the body make up the *peripheral nervous system* (PNS). The spinal cord and brain make up the *central nervous system* (CNS).

Types of neuron

Sensory neuron (leads from receptor to CNS)

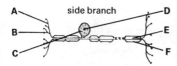

Intermediate neuron (in CNS)

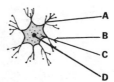

Motor neuron (leads from CNS to effector)

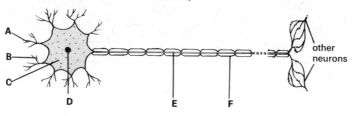

A dendron
B dendrite
C cell body
D nucleus
E axon
F myelin sheath

Impulses These are *electrochemical* messages generated by sodium ions flowing into the axon and potassium ions flowing out of the axon. Impulses can travel at over 100 metres per second.

'All-or-nothing law' A stimulus has to reach a *threshold* intensity before it triggers an impulse. Below this nothing happens; above it, however much larger a stimulus is than the threshold, the same sized impulse results.

Co-ordination

Synapses Where neurons meet they do not touch, but are separated by small gaps called *synapses*. Impulses get across synapses thus:

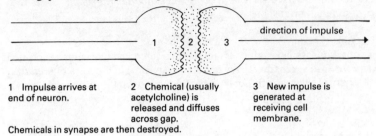

1 Impulse arrives at end of neuron.

2 Chemical (usually acetylcholine) is released and diffuses across gap.

3 New impulse is generated at receiving cell membrane.

Chemicals in synapse are then destroyed.

Nerves of the PNS These are simply bundles of neurons in an insulating nerve sheath.

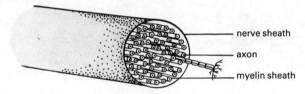

- nerve sheath
- axon
- myelin sheath

Human spinal cord

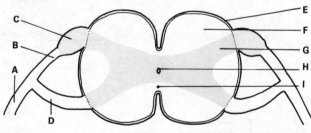

- **A** spinal nerve
- **B** dorsal root
- **C** ganglion
- **D** ventral root
- **E** meninges
- **F** white matter
- **G** grey matter
- **H** central canal
- **I** blood vessel

Co-ordination

Human brain

This has a mass of 1.3–1.4 kg and contains about 10^{10} neurons.

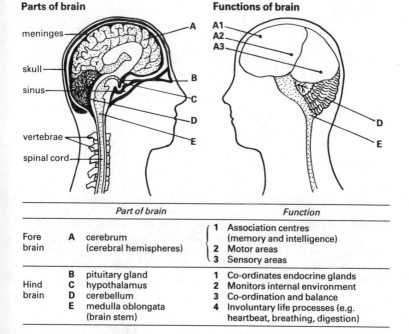

Parts of brain — meninges, skull, sinus, vertebrae, spinal cord; labels A, B, C, D, E.

Functions of brain — A1, A2, A3, D, E.

	Part of brain		Function
Fore brain	A	cerebrum (cerebral hemispheres)	1 Association centres (memory and intelligence) 2 Motor areas 3 Sensory areas
Hind brain	B C D E	pituitary gland hypothalamus cerebellum medulla oblongata (brain stem)	1 Co-ordinates endocrine glands 2 Monitors internal environment 3 Co-ordination and balance 4 Involuntary life processes (e.g. heartbeat, breathing, digestion)

How the parts of the nervous system interact

Simple reflexes These are quick, automatic responses, present at birth and given on receipt of a definite external stimulus. They involve *two* neurons. Examples are:

(*a*) coughing when a foreign body irritates the respiratory system;
(*b*) jerking the knee when stretch receptors are hit;
(*c*) blinking when an object approaches the eye.

Co-ordination

Complex reflexes These too are quick, automatic responses. They are associated with painful stimuli. They involve *three* neurons.

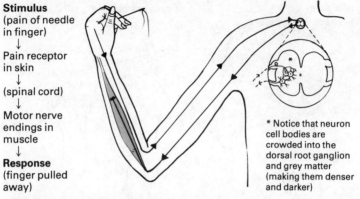

Stimulus
(pain of needle in finger)
↓
Pain receptor in skin
↓
(spinal cord)
↓
Motor nerve endings in muscle
↓
Response
(finger pulled away)

* Notice that neuron cell bodies are crowded into the dorsal root ganglion and grey matter (making them denser and darker)

Total time taken = 1/50th second

Conditioned reflexes These are slower, *learnt* responses. Activities such as riding a bicycle, eating with a knife and fork include many conditioned reflexes, and these involve *association centres* in the *brain*. Conditioned reflexes were first demonstrated by the Russian scientist *Ivan Pavlov* who carried out a famous experiment using dogs.

Pavlov's experiment

1. Dog salivates on sight of food (*simple reflex*)
2. repeated many times — conditioning or training period
3. Dog salivates on sound of bell (*conditioned reflex*)

The dog learnt to associate the sound of the bell with getting food and its salivary glands started working at the sound, even when no food was present.

Intelligent behaviour This is confined to animals with large brains. It includes *problem-solving*. Sensory information goes to the brain and many associations, e.g. with information stored in the memory, may be made before action is taken. This makes reaction time slower.

Endocrine (hormone) system

Hormones are chemical messengers made in ductless *endocrine glands*. They are carried to *target organs* in *blood plasma*.

Human endocrine glands

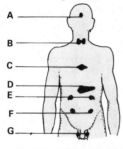

A pituitary
B thyroid
C thymus
D pancreas
   (islets of Langerhans)
E adrenals
F ovaries *or*
G testes

	Hormone(s) made	Effects	Deficiency problems
A	Phyone	Regulates bone growth	Physical and mental retardation
	Gonadotrophin	Regulates menstruation Regulates sperm production	Infertility in adults
	Thyrotrophine (THS)	Controls hormone output of thyroid	(see B)
	Prolactin	Stimulates milk production	Inability to breast feed
B	Thyroxin	Regulates growth and metabolic rates	Cretinism in children Myxoedema in adults
C	Thymus hormone	Delays onset of sexual maturity	Sexual problems in children
D	Insulin	Regulates sugar metabolism	Sugar diabetes
E	Adrenalin	Increases heartbeat, breathing rate, blood glucose level, diverts blood	Addison's disease
	Steroids	Many effects on membranes, etc.	
F	Oestrogen	Controls secondary sex characteristics	Sexual problems in adults (e.g. no menstrual cycle)
	Progesterone	Prepares body for pregnancy	
G	Testosterone	Controls secondary sex characteristics	Sexual problems in adults

THE INDIVIDUAL ORGANISM
3.15 Homeostasis

Homeostasis is the mechanism which brings about a *constant internal environment*. It allows cells, in particular their enzymes, to function under optimum conditions and therefore at maximum efficiency. Homeostasis is an active process and requires energy. It follows the general pattern of three stages: 1 Detection by a *receptor* 2 Message sent via *control centre* 3 Effect by *effector*.

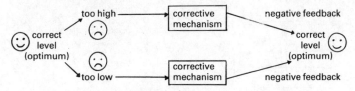

Factors that must be kept constant in humans

Factors	Organ(s) involved	Optimum level
Glucose	Liver and pancreas	1 g per dm^{-3}
Carbon dioxide	Lungs	500 cm^3 per dm^{-3}
Oxygen	Lungs	200 cm^3 per dm^{-3}
Temperature	Skin and liver	37.0°C

(One g per dm^{-3} equals one gram per litre.)

Examples of homeostasis in man

Glucose

Receptor	Control centre	Effector(s)
pancreas	pancreas + adrenals	body cells, liver + muscles

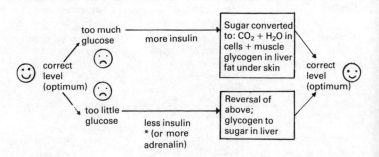

* In **emergencies**, or under nervous or emotional stress, the hormone adrenalin is secreted by the adrenal glands.

Temperature

Receptor	Control centre	Effector(s)
hypothalamus	hypothalamus	skin, lungs, liver muscles

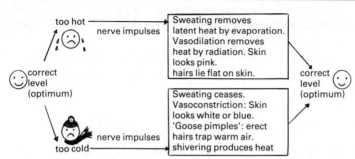

Drugs and their effects

The main groups of drugs are:

(a) **Hallucinogens** (e.g. cannabis, heroin, LSD). These have powerful effects on the nervous system. They are often temporarily pleasurable, but with damaging long-term effects on the body. Most are addictive. *Overdoses* cause *comas* that can be fatal. Sale and unsupervised use of them is *illegal*.

(b) **Analgesics** (e.g. aspirin, paracetamol, morphine, codeine, distalgesic). These are *painkillers*. They are frequently taken for headaches, period pains, etc., despite the hazards of long-term use. They can be addictive.

(c) **Sedatives** (e.g. alcohol, barbiturates, morphine, 'tranquillisers', valium). These are drugs which *relieve anxiety* and *promote relaxation*. They are sometimes called *depressants*. As well as their calming effects, they cause drowsiness. Reactions are slowed and speech is slurred. They can be addictive. Addiction leads to permanent brain and liver damage. *Alcohol* is a particularly common problem, especially dangerous when taken by people driving.

(d) **Stimulants** (e.g. caffeine, cocaine, 'pep-pills', nicotine). These are drugs which *quicken responses*. They can be addictive.

THE INDIVIDUAL ORGANISM
Sample questions

1. Plant and animal cells both possess
   - A nucleus, cytoplasm, wall
   - B membrane, nucleus, vacuole
   - C membrane, nucleus, cytoplasm
   - D chloroplast, nucleus, vacuole

2. The animal cell in this list which has no nucleus is
   - A white blood cell B red blood cell C nerve cell D muscle cell

3. Cells with cilia on them include those
   - A lining nose B lining airducts C lining trachea D all of these

4. Human blood cells are approx. 8 um (8/1000 mm) in diameter. How many could be lined up on a line 1 mm long?
   - A 125 B 250 C 800 D 8000

5. A microscope has a ×5 eyepiece and objectives which magnify ×10 and ×40. Its highest magnification is
   - A ×200 B ×400 C ×1000 D ×2000

6. Enzymes are *not*
   - A sensitive to pH
   - B sensitive to temperature
   - C concerned with respiration
   - D concerned with digestion only

7. Kwashiorkor can be dealt with by giving
   - A antibiotics B extra protein C extra vitamin B D a vaccination

8. Which food type contains most kiloJoules per gram?
   - A sugar B protein C starch D fat

9. Which reagent is used to test for starch?
   - A Benedict's solution
   - B Biuret solution
   - C Iodine solution
   - D Ethanol

10. Beri-beri is caused by a deficiency of
    - A Vitamin C B Iron C Vitamin B D Iodine

11. Carbon dioxide turns
    - A Hydrogen carbonate indicator red
    - B Hydrogen carbonate indicator yellow
    - C Hydrogen carbonate indicator purple
    - D lime water clear

12. Mineral salts are absorbed fastest by plants when they are
    - A transpiring rapidly
    - B flowering
    - C in the dark
    - D losing their leaves

13. A herbivore's gut has an enlarged
    - A duodenum and ileum
    - B caecum and appendix
    - C colon and rectum
    - D caecum and colon

14. Conditions in the human stomach are
    - A pH 7 B neutral C ph 13 D acidic

15 In anaerobic respiration in humans the products are
  A mainly organic   **B** mainly inorganic
  C made without energy release   **D** exhaled

16 Which of the following is *not* a gas exchange surface?
  A mammal's lung   **B** frog's skin
  C insect's spiracle   **D** bird's air sac

17 The main support in the ear pinna is given by
  A blood pressure **B** bone **C** cartilage **D** turgor pressure

18 Which of the following is not a receptor cell?
  A nerve ending in skin   **B** the ear drum
  C rod in retina   **D** proprioceptor in vein

19 Reflex actions do *not* include
  A chewing **B** sneezing **C** coughing **D** blinking

20 Adrenalin secretion increases
  A blood flow to skin   **B** the amount of urine made
  C growth rate   **D** blood sugar level

21 The graph below shows the activity of three human digestive enzymes working within separate ranges of pH

  (*a*) Label neutral, acid and alkali ranges of pH (1)
  (*b*) Label curves relating to trypsin, salivary amylase and pepsin. (1)

22 (*a*) Study the three cells below and list *four* differences between cell **A** and cell **B**. (4)

  (*b*) Animal cells are often specialised. Their shape suits their function.
   (i) State two differences between cell **A** and cell **C**. (2)
   (ii) Explain how the differences in (i) are related to the functions of cell **C**. (2)

(c) (i) What is the function of a chloroplast? (1)
(ii) Where might cells containing many chloroplasts occur? (2)
(d) (i) What is the function of a mitochondrion? (2)
(ii) Where might cells containing many mitochondria occur? (2)

23 The diagram below represents a sealed aquarium kept in natural light.

All living things must obtain energy to survive.
(a) What is the source of energy for green plants? (2)
(b) What is the immediate source of energy for the herbivores? (1)
(c) What is the immediate source of energy for the carnivores? (1)
(d) What changes would you expect in the aquarium if all the animal life were removed and the aquarium re-sealed? (3)
(e) Would it be possible for life to continue indefinitely in the sealed aquarium as shown in the diagram? Explain your answer. (3)

24 The diagram below shows the volumes in cm^3 of human lungs in various stages of inspiration and expiration.

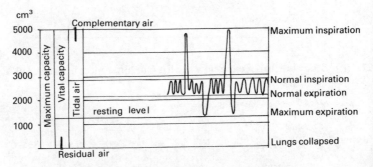

(a) What volume of air is taken in at a normal breath? (1)
(b) A person breathes out fully. What volume of air is taken in if the person then breathes in as fully as possible? (1)
(c) What is meant by residual air? (1)
(d) What is meant by complementary air? (1)
(e) What volume of air is contained at resting level? (1)

25 The diagram below represents photosynthesis in a flowering plant.

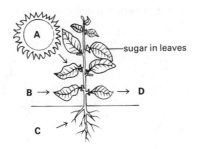

(a) What are the requirements for photosynthesis represented by **A**, **B** and **C**? (3)
(b) Sugar is a main product of photosynthesis, but what is the by-product represented by **D**? (1)
(c) A plant with variegated leaves can be used to show that chlorophyll is needed for photosynthesis. What makes a variegated leaf suitable for this purpose? (1)
(d) Examine the diagram below. This shows details of two parts of a leaf involved in the passage of raw materials for photosynthesis.

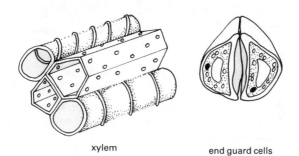

xylem end guard cells

  (i) How is xylem suited to its function in transporting water? (1)
  (ii) How is the stoma opened? (1)
(e) Plants grown without magnesium have yellow leaves. Why do plants require magnesium? (1)

26 (a) What are the characteristics of a good gas exchange surface? (5)
(b) How is gas exchange achieved in
  (i) human (ii) frog (iii) bird (15)

DEVELOPMENT AND REPRODUCTION
4.1 Chromosomes and cell division

Chromosomes; sites of genetic information

Genetic information is situated in the *chromosomes* of the cell nucleus. Individual chromosomes are visible just before and during cell division, if suitably stained. They are made of a unique chemical *deoxyribonucleic acid* or DNA. The DNA molecule is very large and is roughly the shape of a long ladder that has been twisted into a spiral. The two halves of the ladder are very similar. Each has a twisted backbone of *sugar* molecules, linked together by *phosphate* molecules, each carrying a particular chemical *base*. When two bases pair, they form the rungs of the ladder and hold the two halves together as long as necessary.

A small part of the chromosome, just enough DNA to organise the production of a single specific protein, is called a *gene*. Each gene has its own characteristic arrangement of bases which makes it different from all others.

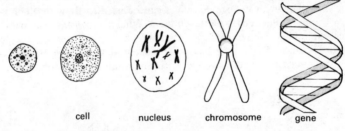

cell　　　nucleus　　　chromosome　　　gene

The genes, and hence the chromosomes, carry:

1 instructions for controlling normal activities in the cell, and
2 instructions for producing a new cell at cell division.

In the body of a multicellular organism, nearly all cells have the same number of chromosomes. This is called the *diploid* number of chromosomes and is represented by 2n. In man, 2n = 46 and in the fruit fly, 2n = 8.

The exceptions are the *gametes* of sexually reproducing organisms (and the cells which give rise to them). These contain half the diploid number of chromosomes. This is called the *haploid* number of chromosomes and is represented by n. In man, $n = 23$, and in the fruit fly, $n = 4$.

In each body cell nucleus, a chromosome can be matched with another of very similar appearance. These form a *homologous pair*, and in a sexually reproducing organism one of the pair originates from the mother and one from the father.

Cell division

Any division of a cell must include a division of the chromosomes and hence DNA, if the resulting *daughter cells* are to carry appropriate instructions to survive.

Division of DNA
(*a*) The DNA molecules 'unzip' between the base pairs.
(*b*) New bases floating in the nucleus attach to each half of the ladder.
(*c*) A new backbone of sugar and phosphate molecules attaches to the base pair.

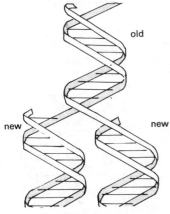

So the original molecule has acted as a *template*, making half of each of the new molecules. The outcome is the *doubling* of each chromosome into two strands, the *chromatids* which stay together because one part, the *centromere*, does not immediately divide.

A comparison of the two types of cell division

Mitosis	*Meiosis*
Two daughter cells formed in each cycle.	Four daughter cells formed in each cycle.
Daughter cells are identical to parent cell and each other.	Daughter cells are different from parent cell and each other.
Amount of genetic 'information' stays the same, i.e. *diploid* ($2n$).	Amount of genetic 'information' is halved, i.e. *haploid* (n).
One cell division.	Two cell divisions.
Occurs in body cells, e.g. animals' skin and bone plants' apical meristems (root and shoot tips).	Occurs only in reproductive cells that give rise to gametes, e.g. animals' ovaries and testes, plants' sporangia, ovules and anthers.

Mitosis

This type of cell division is associated with *growth* and *asexual reproduction*. It is a continuous process, but different stages can be recognised. (For the sake of clarity, the nucleus of the cell below is drawn disproportionately large. It is a fruit fly cell with a diploid number of 8.)

Interphase
Cell is growing.
DNA divides so
each chromosome =
2 chromatids.

Prophase
8 chromatids visible.
Centriole divides
and moves to poles.
Nuclear membrane
disappears.

Metaphase
Spindle forms between
centriole.
Chromosomes attach to
spindle at centromeres.

Anaphase
Chromatids
separate.
Centromeres lead
to the poles.

Telophase
Spindle disappears.
Nuclei reform.
Cytoplasm splits.

The two new cells
grow to full size.
The DNA of the
chromosomes then
divides again.

Chromosomes and cell division 134

Meiosis

This type of cell division is associated with sexual reproduction in *gamete formation*. It is sometimes called *reduction division*, because the second half of the process halves the amount of DNA in a nucleus. Again, it is a continuous process but different stages can be recognised. These are not the same as in mitosis. (A cell with just one pair of chromosomes is used below to outline events.)

Whole pair of chromosomes come together.

Each chromosome duplicates to give two chromatids joined by centromeres.

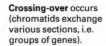

Crossing-over occurs (chromatids exchange various sections, i.e. groups of genes).

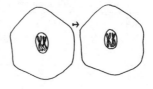

Chromosomes, now slightly altered, move to opposite poles. Two different daughter cells form.

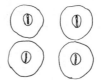

Each daughter cell now divides, as in mitosis.

Meiosis generates variation by **crossing-over** and **random assortment** of chromosomes.

DEVELOPMENT AND REPRODUCTION
4.2 Growth and life cycles

What is growth?

Growth of an individual organism can be defined as a *permanent increase in size*. (Temporary increases due to phenomena such as pregnancy or uptake of water are not growth.)

Measuring growth

A number of different measurements can be made and used to describe growth. Common ones are wet (fresh) mass or weight, dry mass or weight, length, width and number of cells. Which is chosen will depend on the type of organism and the purpose of the study. The units used must be appropriate too. Consider the following.

Parameter	Main advantages	Main disadvantages
Fresh weight	Easy (surface water removed and organism weighed at intervals). Does not harm the organism. Suitable for most organisms.	Influenced by changes in water content (e.g. drinking, wilting).
Dry weight	More accurate.	Time-consuming. Involves death of organism (kept at 100°C until all water is removed).
Length	Easy.	Ignores growth in all but one direction.
Width	Easy.	Ignores growth in all but one direction.
Number of cells	Accurate.	Difficult. Very limited use (e.g. for *Spirogyra* tissue cultures, etc.). Involves using microscope.

Other parameters Sometimes biologists are concerned with only one part of an organism. This is especially true for crop plants where only flowers (e.g. roses), fruit (e.g. apples) or leaves (e.g. lettuce) are important to the grower. In these cases, the parts which form a product for sale may be considered separately.

Growth and life cycles 136

How growth occurs

Growth is a characteristic of all living things and reflects their way of life. For example, animals that need to move to get food keep a compact shape as they grow, whereas land plants remain in one place and have a spreading shape to maximise absorption of water, nutrients and sunlight energy. Growth involves:

(*a*) formation of cytoplasm,
(*b*) cell division (mitosis),
(*c*) vacuolation (plants only),
(*d*) differentiation.

Examples of growth that you might be expected to know are: seed germination, root growth, stem growth, twig growth, human growth.

Seed germination

The **most** important factors needed for germination are:

Moisture to activate enzymes to release food store (enters through the *micropyle*);
Oxygen for aerobic respiration to supply energy;
Suitable level of
Temperature to speed enzyme action and respiration.

Note: As soon as green cotyledons or leaves appear, photosynthesis can begin. Until this point, very few seeds require light.

Practical test to demonstrate factors needed for germination

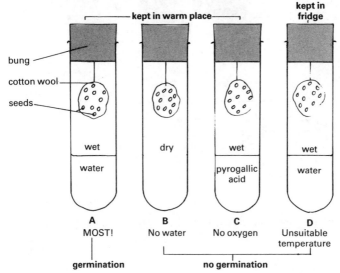

Make sure that you can follow these two examples and know what each term means. You may, however, prefer to learn details of species you have studied.

Wheat (*Triticum*), a monocotyledon, demonstrates hypogeal germination.

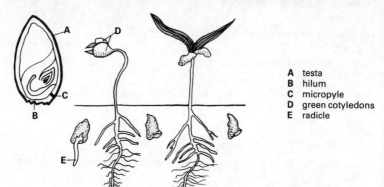

A testa
B hilum
C micropyle
D green cotyledons
E radicle

The whole seed is pushed out of the soil by the *plumule*. The *cotyledon* is first to photosynthesise.

Sunflower (*Helianthus*), a dicotyledon, demonstrates epigeal germination.

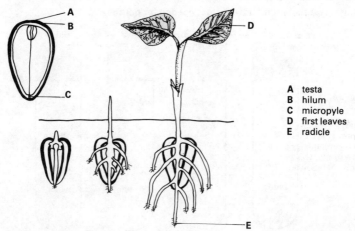

A testa
B hilum
C micropyle
D first leaves
E radicle

The cotyledons remain below ground, so *leaves* are first to photosynthesise.

Growth and life cycles   138

Root growth (*demonstrates all the processes of growth*)

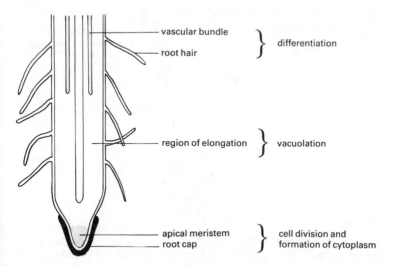

(Cross-section of root is shown on p. 83.)

Stem growth (*in woody plants follows a fixed sequence*)

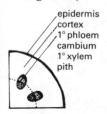

Primary (1°) phloem and primary (1°) xylem form behind the shoot tip in a similar way to the root tip (above).

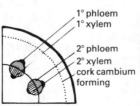

The cambium divides to make secondary (2°) phloem and secondary (2°) xylem, which becomes *wood*.

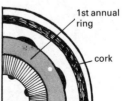

Cork cambium makes cork to waterproof stem. First annual ring becomes visible in xylem which grows faster in summer and slower in winter.

Twig growth

A winter twig (e.g. Horse Chestnut)

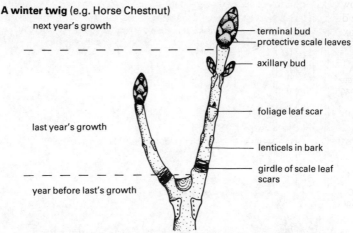

Late summer growth is protected inside *buds* during the cold winter. In spring, buds grow and leave *girdle scars* where rings of scale leaves fall off. *Leaf scars* are left in autumn.

Human growth

Different parts of the human body grow at different rates. This is called *allometric growth*.

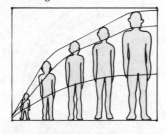

Growth from birth to adulthood proceeds along a curve

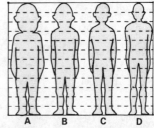

- **A** Newborn baby: head one-quarter of body length
- **B** Infant
- **C** Adolescent
- **D** Adult: head one-eighth of body length

Patterns of growth

These are usually represented by graphs. If curves are smooth, i.e. the organism is gradually increasing in size and mass, growth is said to be *continuous*. Where curves are stepped, i.e. there are sudden changes in the organism's size and mass, growth is said to be *discontinuous*.

Growth and life cycles 140

Study the following graphs and decide which show continuous and which show discontinuous growth. Try to identify the organism for each. Notice the differences caused by choice of measurements (**A** and **B**) and by choice of axes (**C** and **D**).

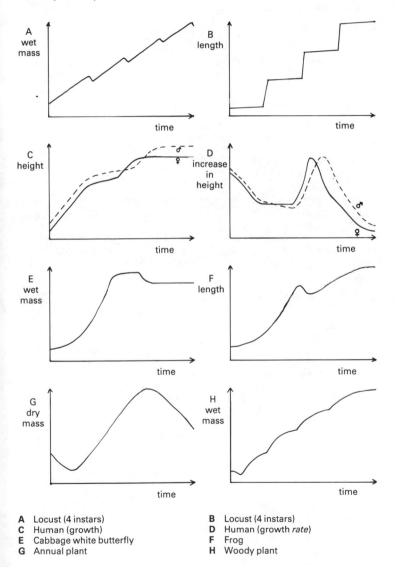

- **A** Locust (4 instars)
- **B** Locust (4 instars)
- **C** Human (growth)
- **D** Human (growth *rate*)
- **E** Cabbage white butterfly
- **F** Frog
- **G** Annual plant
- **H** Woody plant

Growth and life cycles

Metamorphosis

Metamorphosis is a change in the body form and way of life of an organism between its egg and adult stages. Most examples belong to the insect and amphibian groups. The different forms at different stages are sometimes called *instars*.

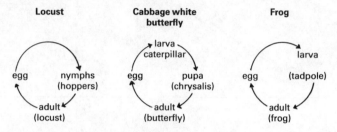

An adult insect is sometimes called an *imago*.

Control of growth

	Controlling factors	Effects/comments
For plants	**External**	
	Temperature	Warmth promotes respiration.
	Light intensity	Increasing light intensity promotes photosynthesis.
	Minerals + water availability	May include artificial fertilisers and irrigation.
	Internal	
	Hormones	e.g. auxins, giberellins, kinins.
	Inherited features	e.g. tall and dwarf varieties.
For animals	**External**	
	Temperature	Affects cold-blooded animals. Warmth promotes metabolism.
	Food availability	A balanced diet is needed. Proteins are particularly important
	Internal	
	Hormones	e.g. phyone: pituitary growth hormone; thyroxin: from thyroid controls growth rate; oestrogen and testosterone: sex hormones; control female and male sex characteristics.
	Inherited features	e.g. tall and dwarf varieties.

Life cycles

Individuals of a species have a fixed *life expectancy*. Barring accidents or disease, they live this long but then die. If they were not replaced by *reproduction*, different species would die out. Each has a set *life cycle*.

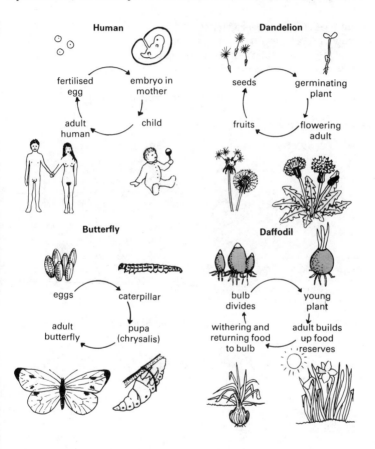

All plants and animals make more of their own kind, i.e. humans have babies, horses have foals, cats have kittens and oak trees have acorns, which grow into oak trees. There are two types of reproduction. *Asexual* and *sexual*. Some organisms demonstrate both types.

DEVELOPMENT AND REPRODUCTION
4.3 Asexual reproduction

Asexual reproduction involves *one parent* only. *All offspring are identical* to the parent and each other, and are called *clones*. It occurs in many plants and simple animals.

Advantages of asexual reproduction
1. Rapid process, involving only mitosis.
2. Provides exact replication of strong individuals.
3. The haphazard processes of pollination, seed dispersal and germination is avoided.
4. In an unchanging environment, all offspring may survive.

Examples of asexual reproduction

Fission (e.g. *Amoeba*, bacteria)   **Budding** (e.g. *Hydra*, yeast)

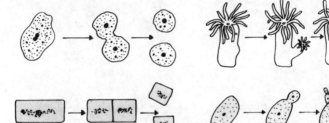

Spores (e.g. mould fungi, mosses)   **Parthenogenesis** (e.g. greenfly, bees)

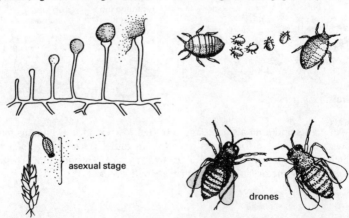

Asexual reproduction 144

Vegetative propagation

Bulbs (e.g. daffodil)

Corms (e.g. crocus)

Stem tubers (e.g. potato)

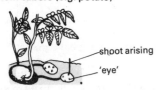

shoot arising
'eye'

Rhizomes (e.g. mint)

stem

Runners (e.g. strawberry)

runner

Cuttings (e.g. geranium (*Pelargonium*))

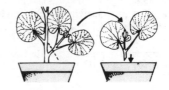

Plantlets (e.g. *Bryophyllum*)

small plants arise on leaf margin

Cloning: tissue culture (e.g. carrot)

culture medium 'in vitro'

Grafting (e.g. apple)

Buds

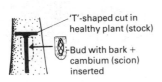

'T'-shaped cut in healthy plant (stock)

Bud with bark + cambium (scion) inserted

Stems

stem of scion (shape + cambium matched)
stem of stock

grafts sealed with wax and bound

DEVELOPMENT AND REPRODUCTION
4.4 Sexual reproduction

This usually involves *two parents*, of different sexes, one *male* (♂) and one *female* (♀). These produce special *sex cells*, called *gametes*. They are made by *meiosis* in the *reproductive organs*, called *gonads*. Offspring are different from the parents and each other, deriving half their genes from each parent. Sexual reproduction occurs in almost all plants and animals.

Simple organisms such as *Mucor* and *Spirogyra* have different *strains* but not easily-distinguishable sexes. A few organisms are *hermaphroditic*; they have both male and female gonads.

Advantages of sexual reproduction
1. Produces new varieties, allowing evolution of the species.
2. Allows invasion of new territory by dispersal and prevents competition between parents and offspring.
3. In a changing environment, *variation* helps survival. (When all offspring are identical, a change in the environment that kills one is likely to kill them all.)

Gametes
These are produced by *meiosis* and therefore have half the usual number of chromosomes, i.e. they are *haploid* (n).

Examples of gametes in plants

Ovule (♀)

Pollen (♂)

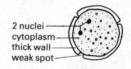

- 2 nuclei
- cytoplasm
- thick wall
- weak spot

- integuments
- ovule (embryosac)
- cytoplasm
- nucleus

Examples of gametes in animals

Ovum (♀)

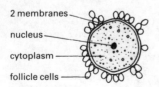

- 2 membranes
- nucleus
- cytoplasm
- follicle cells

Sperm (♂)

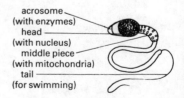

- acrosome (with enzymes)
- head (with nucleus)
- middle piece (with mitochondria)
- tail (for swimming)

Sexual reproduction 146

Fertilisation

This is the *fusion* of the nuclei of a male gamete and a female gamete to form a *zygote*. The *diploid* chromosome number is restored.

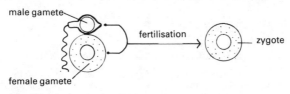

External fertilisation occurs outside the body of the female
Internal fertilisation occurs inside the body of the female.

There is a trend towards internal fertilisation in higher plants and animals. External fertilisation requires water as a medium in which sperms can swim to female gametes (and later to stop the developing zygote from drying out). It is therefore found in aquatic and primitive land organisms (the mosses and liverworts of the plant kingdom and the amphibians of the animal kingdom).

External fertilisation tends to be more haphazard and involves great wastage of gametes. Hence enormous numbers are produced. (It is for this reason also that flowering plants relying on wind to transmit their pollen make larger quantities).

Internal fertilisation in animals requires a male and a female to come together so that gametes reach each other. They mate or *copulate*. Special organs (e.g. the penis) have evolved, together with suitable *behaviour* to encourage pairing. A great deal of *energy* is used to bring about fertilisation to ensure *breeding success*.

147 *Sexual reproduction*

Sexual reproduction in plants

A In simple plants: *conjugation* (e.g. *Spirogyra, Mucor*)

In *Spirogyra* a conjugation tube forms from cells of two filaments

Nucleus (with cytoplasm) of one cell moves and fertilises the other to make a zygote

Zygote secretes a spore coat to give a tough zygospore

Zygospore germinates into a new filament in spring

In *Mucor*, gametangia full of nuclei grow towards each other

Gametangia fuse and fertilisation occurs in pairs of nuclei

A thick spore wall forms round zygotes

Zygospore germinates into a new sporangium

B In higher plants: *pollination, fertilisation, seed and fruit formation*, and *seed dispersal*, e.g. flowering plants (angiosperms).

Pollination This is the transfer of *pollen* from an *anther* to a *stigma*. (Check that you remember flower parts, page 5.)

  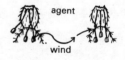

Self-pollination
The transfer of pollen from anther to stigma of the *same plant*.

Cross-pollination
The transfer of pollen from anther to stigma of a *different plant of the same species*.

Sexual reproduction

Cross-pollination has the advantage of producing much greater *variation* in offspring. Mechanisms have evolved to promote this, namely:

(a) *separate sex plants*, e.g. squash;
(b) *special structures*, e.g. primrose (pin-eyed and thrum-eyed);
(c) *protandry*, e.g. dandelion (stamens ripen before carpels);
(d) *protogyny*, e.g. plantain (carpels ripen before stamens);
(e) *incompatibility*, e.g. primrose (the stigma produces chemicals preventing the plant's own pollen from germinating).

Insect and wind pollination Plants which reproduce by these methods are identifiable as follows:

Insect-pollinated flowers have:	*Wind-pollinated* flowers have:
Large, conspicuous coloured petals;	Small, inconspicuous greenish petals;
Scent;	No scent;
Nectaries;	No nectaries;
Rigid, protected flower parts;	Flexible, exposed flower parts;
Small anthers producing large sticky pollen;	Large anthers producing small dry pollen;
Small stigmas and unbranched styles.	Stigma and style large and branched.
e.g. Foxglove Lily Iris Eyebright Strawberry	e.g. Barley Lamb's tongue (plantain) Oats Wheat Sedges

Fertilisation Compatible (suitable) pollen grains germinate on the stigma. They form *pollen tubes* and grow down the style, entering the ovary via the *micropyle*.

Pea (insect-pollinated)

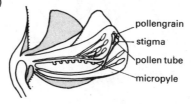

One of the nuclei fuses with the ovule. The other nucleus fuses to form the first food-storing cell.

Sexual reproduction

Seed and fruit formation The diagrams below show parts of a pea plant. Follow the fates of flower parts as fruit and seed are formed. Parts that attracted insects are of no further use and wither.

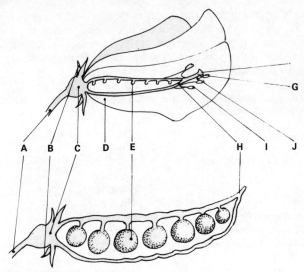

- **A** flower/fruit stalk
- **B** sepal/sepal
- **C** receptacle/receptacle
- **D** petals/–
- **E** ovule/seed
- **F** stigma/–
- **G** style/–
- **H** ovary/fruit (pod)
- **I** anther/–
- **J** filament/–

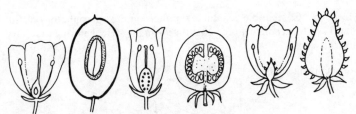

Damson   Tomato   Strawberry

Fruits have three functions. They:

(a) Protect the seed.
(b) Disperse the seed.
(c) May provide nutrients for germination.

Sexual reproduction 150

Dispersal of fruits Dispersal (spreading) of seeds prevents overcrowding of seedlings and helps invasion of new ground. There are *four* methods of dispersal.

By animals

burdock

blackberry

hooks to fur, feather or clothes

eaten and passes through gut unharmed

By wind

dandelion

sycamore

large surface area (parachute of hairs)

large surface area (wings to twist fruit and slow fall)

By water

coconut

water lily

air in fruit gives buoyancy

oil in fruit gives buoyancy

By mechanical means (self-dispersal)

pea

geranium

pod dries and strain builds up until pod splits and twists

capsule dries and explodes

Human sexual reproduction

Male reproductive organs

Front view (section)

Side view (half body, with penis erect)

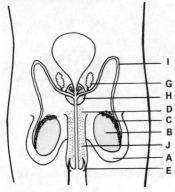
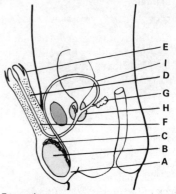

- **A** scrotum
- **B** testis
- **C** epididymis
- **D** penis
- **E** foreskin
- **F** urethra
- **G** seminal vesicle
- **H** prostate gland
- **I** sperm duct (vas deferens)
- **J** erectile tissue

Female reproductive organs

Front view (section)

Side view (half body)

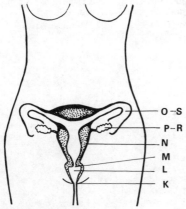
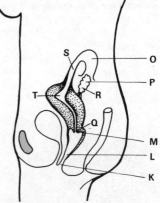

- **K** vulva
- **L** vagina
- **M** cervix
- **N** uterus wall
- **O** oviduct (Fallopian tube)
- **P** ovary
- **Q** where sperms are deposited
- **R** where ovum is released
- **S** where fertilisation occurs
- **T** where embryo implants and develops

Functions of human reproductive organs

(These should be checked against the labels on the diagrams.)

Male
Production of sperms	**B**
Contain the testes and maintain them below 37°C (necessary for sperm development and survival)	**A**
Tubes taking sperms to urethra by peristalsis	**I**
Coiled tube for sperm storage until ejaculation	**C**
For insertion into the vagina for sperm release	**D**
Secretion of seminal fluid	**G, H**
Carrying sperm for reproduction and urine for excretion	**F**

Female
Production of eggs (ova)	**P**
For receiving penis during copulation. Later becomes birth canal	**L**
Neck of the uterus, allowing passage of sperm and later baby	**M**
Receives zygote and protects embryo	**N**
Site of fertilisation. Directs ovum or zygote towards uterus	**O**
Opening of vagina	**K**

Puberty in boys (11–16 years)
1 Testes grow and secrete *testosterone*.
2 Testosterone triggers development of secondary sex characteristics:
   (*a*) voice breaks as larynx grows,
   (*b*) muscles give male shape,
   (*c*) beard and pubic hair grow.
3 Penis enlarges.

Structure of the testis

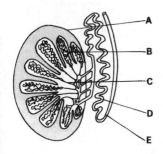

A outer wall
B sperm-producing tubules
C collecting ducts
D epididymis
E sperm duct

Puberty in girls (9–16 years)
1 Ovaries grow and secrete *oestrogen*.
2 Oestrogen triggers development of secondary sex characteristics:
   (*a*) breast development,
   (*b*) fat under skin gives female shape,
   (*c*) underarm and pubic hair grow.
3 Menstruation (periods) starts.

Menstruation The reproductive cycle in human females is called the *menstrual (monthly) cycle*. It begins at *puberty* and repeats itself at monthly intervals until *menopause* (45–55 years), except during pregnancy when the yellow body and placenta secrete hormones to delay the cycle.

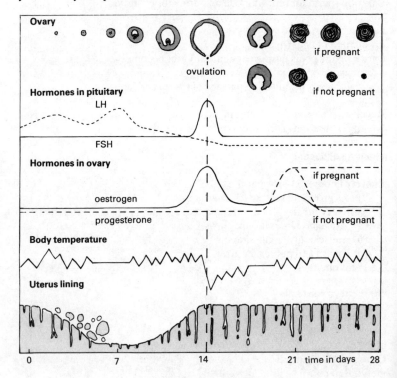

Structure of the ovary

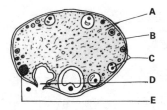

A blood vessels
B capsule
C maturing follicles
D Graafian follicle bursting
E corpus luteum (yellow body)

Copulation (sexual intercourse)

The erectile tissue of the penis fills with blood and can then be inserted into the vagina, helped by secretions from the female. Stimulation of sensitive cells at the tip of the penis and in front of the vagina (the clitoris) give both partners pleasure and in the male set off a reflex action resulting in *ejaculation*. Up to 400 000 000 sperms are squeezed (by peristalsis) from the epididymis, via the sperm duct, collecting seminal vesicle and prostate gland secretion on the way. The *semen* passes into the urethra and is deposited at the top of the vagina. Recently '*in vitro*' fertilisation has been developed. Where a couple cannot conceive, an ovum can be fertilised by a sperm outside the body, and then implanted in the mother's uterus, giving rise to a so-called 'test-tube baby'.

Human fertilisation

Many sperms arrive at ovum. One contacts it. This may be 2–3 days after intercourse.

Enzymes from acrosome break down membranes. Sperm head enters and tail is left behind.

The outer membrane thickens to prevent more sperm entering. Nuclei fuse.

Conception and contraception

Conception is the fertilisation of an ovum by a sperm and the *implantation* of the resulting *zygote* in the uterus lining.

Contraception is a practice or device that prevents this happening. (Most methods of contraception are outlined on page 37.)

Pregnancy (gestation)

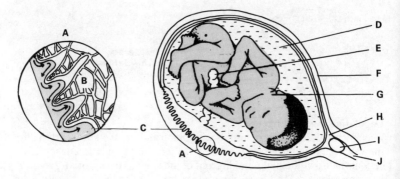

- **A** placenta
- **B** capillaries
- **C** membranes
- **D** amnion (cushions from shock)
- **E** umbilical cord
- **F** uterus wall (strongest muscle in body)
- **G** baby (9 months after conception)
- **H** cervix
- **I** plug of mucus
- **J** vagina (birth canal)

Role of placenta The placenta allows substances to *diffuse* between the mother's and baby's blood. It prevents their blood mixing so that:

(a) different blood groups do not clot,
(b) bacteria cannot pass from mother to baby,
(c) mother's blood pressure does not damage baby's blood vessels.

Substances passing from mother to developing baby	Substances passing from developing baby to mother
Oxygen	Carbon dioxide
Dissolved food (glucose, amino acids, vitamins, minerals, fatty acids and glycerol)	Nitrogenous waste (urea)

Antenatal care (during pregnancy)

Remember 'Doctor, doctor'! He or she will give a pregnant woman regular and free check-ups, and any prescription necessary, and '**Dr, Dr**' will remind you of special considerations, i.e.

DRugs may pass across the placenta and harm the baby, and should be avoided (e.g. alcohol damages nerves and liver);
Diet must take account of the growing baby's needs;
Rest stops swelling ankles. No vigorous exercise or heavy lifting should be undertaken.

Sexual reproduction 156

Birth (parturition)
This is best considered in three stages:

1 **Labour:** *contractions* of the uterus, triggered by hormones (e.g. oxytocin), push the baby's head against the cervix and gradually dilate (open) it. The amnion bursts and the amniotic fluid escapes through the vagina. This is called the *'breaking of the waters'*.
2 **Delivery:** the baby is usually born by moving head-first down the birth canal. The rest of the body follows easily, as the head is the largest part. The umbilical cord is clipped and cut. The average mass of a newborn baby is 3 kg. If a baby is born feet-first, it is termed a *breech birth*.
3 **Afterbirth:** contractions continue and force the rest of the umbilical cord and the placenta out of the uterus.

Difficult deliveries may include the use of *forceps* (large tongs with spoon-shaped ends) to guide the baby's head or, in rarer cases, a *Caesarian section* delivers the baby through an incision (surgical cut) through the abdomen and uterus walls, under anaesthetic.

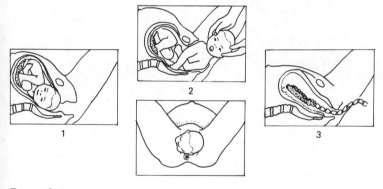

Parental care
Most human mothers, like all mammals, *breast feed* or *suckle* their young on milk made in *mammary glands*. This is stimulated at birth by the hormone *prolactin*. Milk must contain nearly all essential food types, as it is the baby's only food until it is *weaned* or gradually introduced to solid foods. Some mothers prefer to *bottle feed* babies with a powdered food based on cow's milk. In addition, human parents provide warmth, shelter, education and much more besides for their offspring for many years.

Multiple births

About one in a hundred pregnancies result in the birth of *twins*. Twins can be of two types:

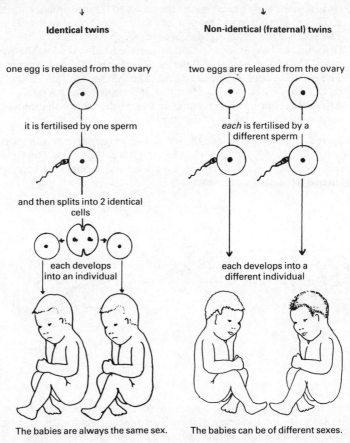

With the increased use of *fertility drugs*, the incidence of multiple births has increased too. The largest set of which all the babies survived is six (*sextuplets*).

Comparison of vertebrate reproduction

Fish (e.g. trout)	Amphibian (e.g. frog)	Reptile (e.g. grass snake)	Bird (e.g. robin)	Mammal (e.g. cat)
Lives in water	Lives in water and on land	Lives on land (and in water)	Lives on land (and in air)	Lives on land
External fertilisation	External fertilisation	Internal fertilisation	Internal fertilisation	Internal fertilisation
Copulation very rare and seasonal	Copulation common and seasonal	Copulation usual and seasonal	Copulation always and seasonal	Copulation always, usually seasonal (oestrus), or artificial insemination
Courtship rare	Courtship simple	Courtship simple	Courtship complex	Courtship complex
No. of gametes enormous	No. of gametes very many	No. of gametes relatively few	No. of gametes relatively few	No. of gametes relatively few
Small amount of yolk in egg	Small amount of yolk in egg	Large amount of yolk in egg	Large amount of yolk in egg	Little or no yolk in egg, placenta
Embryo protected by jelly	Embryo protected by jelly	Embryo protected by leathery shell and buried in warm, wet sand	Embryo protected by calcareous shell and always incubated by parents	Embryo protected by female's body and uterus walls
Parental care usually absent (rare exceptions are seahorse & stickleback)	Parental care of eggs found rarely	Parental care of eggs and young occasional	Parental care intensive and complex, feeding lasting several weeks	Parental care intensive and lasting – young suckled, protected and taught
Survival rate of offspring to adulthood low	Survival rate of offspring to aduldthood low	Survival rate of offspring to adulthood fair	Survival rate of offspring to adulthood good	Survival rate of offspring to adulthood very good

DEVELOPMENT AND REPRODUCTION
4.5 Inheritance

4

Every normal cell in an organism has two sets of chromosomes which can be arranged in homologous pairs. As a gene is a small piece of a chromosome, it follows that every cell has two sets of genes. For each characteristic transmitted there are two genes per cell at identical positions (*loci*) on two homologous chromosomes. These are called *alleles*. The particular genes that an organism has for each characteristic are described as its *genotype*. The detectable features that these give the organism are called its *phenotype*.

Usually one allele 'takes control' and shows itself in the phenotype. It is known as *dominant*. The other gene is usually overshadowed and is known as *recessive*. Letters are used as symbols for genes; dominant ones are written in capital letters, e.g. **A** and **B**, while recessive ones are written in small letters, e.g **a** and **b**. Occasionally alleles are of equal dominance and both contribute to the phenotype. This is known as *co-dominance* or *incomplete dominance*. (Examples are petal colour in Snapdragons and the ABO blood groups in humans.)

The diagram below represents part of a pair of homologous chromosomes before replication at interphase. DNA portions **A**, **B**, **C** and **D** each represent genes in the way described above.

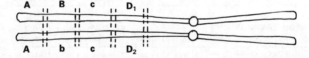

Homozygous (pure-bred): the two alleles in a cell are the same (e.g. **AA**, **aa**).
Heterozygous (hybrid): the two alleles in a cell are different (e.g. **Aa**).

These chromosomes show that this organism has the following genotypes and phenotypes:

Genotype	Description	Phenotype	Genes which show up
AA	homozygous dominant	dominant	**A**
Bb	heterozygous	dominant	**B**
cc	homozygous recessive	recessive	**c**
D_1D_2	heterozygous co-dominant	co-dominant	$D_1 + D_2$

Inheritance 160

When gametes are produced, only half the genetic information (one gene allele) goes into each gamete. All the gametes produced by individuals with genotype **AA** will carry the **A** gene allele as shown below:

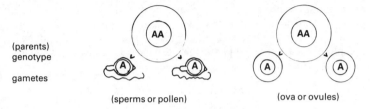

(parents) genotype

gametes

(sperms or pollen)   (ova or ovules)

If these two individuals were crossed, all offspring would have genotype **AA** because all gametes from both parents would carry an **A** gene allele.

If an individual has genotype **Aa** then half its gametes will carry allele **A** and half will carry allele **a** (see below).

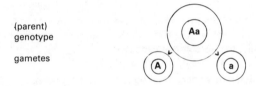

(parent) genotype

gametes

You need to understand this before you can attempt any genetics problem. Check also that you know the following definitions and conventions.

Cross: a mating between a male and a female to produce offspring.
Alleles: genes which control the same characteristic.

P is used as a symbol for the parents of a cross (parental generation).
F_1 is used as a symbol for the offspring of a cross (first filial generation).
F_2 is used as a symbol for the offspring of the F_1 generation (second filial generation). In higher animals, the F_1 generation seldom cross with each other.

Single-factor inheritance: monohybrid cross

Working out genetic crosses is easy, but many students make it difficult for themselves! You need a method to follow each time. There are two common ones in use and the one shown below on the right, the *Punnett square*, is more popular. Whichever you use, always put the same parent on the left. (Examining Groups differ in which they choose.) In the following examples the *male* parent is put on the left-hand side. Using the example on the previous page:

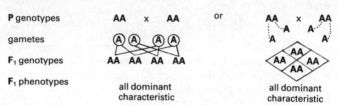

There are three more patterns of cross that you may encounter. Each of these is outlined below. Go through the examples until you are sure that you can predict the outcome of each cross.

Example 1

Let **T** be the dominant gene for tongue-rolling in humans.
Let **t** be the recessive gene for non-tongue-rolling in humans.

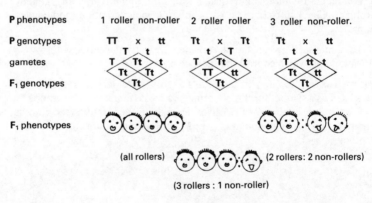

It is impossible to tell from the outside which 'rollers' are **Tt** and which **TT**

A *backcross* (crossing with a homozygous recessive e.g. **tt**) reveals that the other parent was heterozygous. Compare the result with 1.

Inheritance

You carry out exactly the same procedure for a case of *co-dominance*, only this time both genes may show in the phenotype.

Example 2

Let **R** be the co-dominant gene for red petals in Snapdragons.
Let **W** be the co-dominant gene for white petals in Snapdragons.

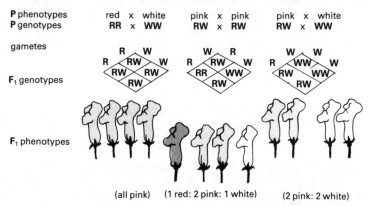

(all pink)	(1 red: 2 pink: 1 white)	(2 pink: 2 white)

From these examples you should see that *which* of the parents' gametes fuse is a matter of *chance*, but there will always be one from the female and one from the male. There are also various *ratios* of offspring that we can *expect* from genetic crosses. If parents had huge numbers of offspring these ratios would hold exactly true, but more often parents have fewer offspring (humans may choose to have just one).

Interpreting experimental results

Even in examination questions, the number of offspring will not be given in *precise* ratio. For example:

What are the likely ratios for these findings for F_1 generations?

(a) Out of 96 fruit flies, 70 had red eyes and 26 had white eyes.
(b) Out of 500 roses, 130 were red petalled, 244 were orange petalled and 126 were yellow petalled.
(c) Out of 9 mice in a litter, 5 had black fur and 4 had white fur.

Answers: (a) 3:1   (b) 1:2:1   (c) 1:1

Mendel's laws of inheritance

Gregor Mendel, an Austrian monk, discovered much of this genetic information by painstaking work over many years, using the ordinary garden pea (*Pisum*). He knew nothing of cell nuclei, chromosomes or genes, and his extraordinary findings led him to be called 'the Father of Genetics'.

Different strains of pea were kept separate and allowed to self-pollinate for several generations. (Little muslin bags were used to cover the flowers.) This meant that experiments started with pure-bred lines and involved controlled cross-pollination of different strains. They were repeated many times and led to the following laws.

First Law of Inheritance

Of a pair of contrasted characters, only one can be represented in a single gamete. In modern language, only one allele for each characteristic enters each gamete produced.

Second Law of Inheritance: Law of Independent Assortment of Characters

Each of a pair of contrasted characters can be combined with each of another pair. In other words, each of a pair of alleles can usually be rearranged to combine with either allele of another pair. (These new combinations explain how a child may have its mother's hair colour and its father's nose shape!)

Mutations

These are changes in cells that can be caused by chance mistakes, or by agents such as chemicals, X-rays and UV light. Most are harmful.

Somatic mutation affects body cells and affects one generation only (e.g. skin cancer caused by UV light).
Genetic mutation affects a gene or whole chromosome and can be passed on to future generations, thereby playing a role in evolution (e.g. melanic form of peppered moth, Down's syndrome in man).

Sex determination

In mammals Sex is inherited by whole chromosomes and not single genes. Males possess two different sex chromosomes (XY), while females possess two similar sex chromosomes (XX). The Y chromosome is dominant to the X chromosome. The male partner thus determines the sex of the offspring. At meiosis the sex chromosomes separate normally: half the sperm carry X and the other half Y. All ova contain an X chromosome.

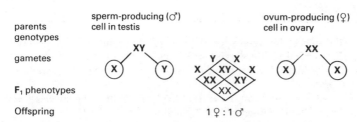

parents genotypes	sperm-producing (♂) cell in testis		ovum-producing (♀) cell in ovary
gametes	XY		XX
F₁ phenotypes			
Offspring		1♀ : 1♂	

In birds The female carries the Y chromosome. Males are XX.

Sex linkage

Some genes carried on sex chromosomes are not concerned with sex characteristics. They are called *sex-linked* genes. Any recessive alleles on the second leg of the female's X chromosome appears in the phenotype of male offspring, as there are no corresponding loci for dominant alleles. (Thus males tend to suffer more from sex-linked complaints, e.g. haemophilia, colour blindness, baldness.)

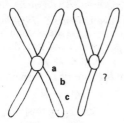

If a recessive gene allele occurs at **a**, **b** or **c**, it will appear in a male child . . .

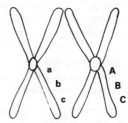

. . . but not in a female child with dominant gene alleles on the second sex chromosome.

DEVELOPMENT AND REPRODUCTION
4.6 Variation

4

The differences between species (*interspecific variation*) can usually be clearly observed as external characteristics. An onion plant and a leek plant are closely related but the two species are easily recognised, as are horses and zebras.

There is also variation between individuals of the same species (*intraspecific variation*). This can be brought about by:

(*a*) the environment;
(*b*) recombination of parents' genes in sexual reproduction;
(*c*) crossing over in meiosis;
(*d*) mutations.

The first gives rise to so-called *acquired characteristics* (e.g. a human's appearance could be altered by a slimming diet or a holiday in the sun!). The other three give rise to *inheritable characteristics*. These are more important, since they equip individuals of a species for survival in different environments and make sure that their offspring have a better chance of survival too.

Intraspecific variation is considered under two headings: *discontinuous variation* and *continuous variation*.

Discontinuous variation
Characteristics are clearly separated into two or more types with no range of intermediates, for example:

ears lobed or unlobed

tongue roller or non-roller

peas wrinkled or round

human blood groups

Such features

(*a*) usually arise as a result of mutation;
(*b*) are usually controlled by one or two pairs of genes;
(*c*) cannot be influenced by environmental factors;
(*d*) are clearly passed on to offspring.

Continuous variation

Characteristics show a continuous range of expression with *very small differences* between individuals, for example:

height

mass

Such features

(*a*) are usually controlled by several genes;
(*b* can be influenced by environmental factors;
(*c*) are not clearly passed on to offspring.

Histograms Continuous variation is usually studied by constructing *histograms* of a sample of a population (e.g. heights of a class of schoolchildren as shown on the left). If a whole population is studied, the steps become very small and a typical histogram approximates to a *normal curve* as shown on the right.

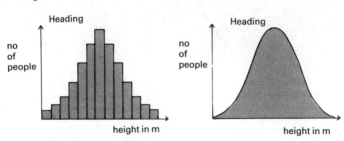

We can look at *three* features of a normal curve:

1. *Mode:* the most common measurement.
2. *Median:* the middle of the range along the *x* axis.
3. *Mean:* the average measurement $=\dfrac{\text{sum of all individual measurements}}{\text{number of individuals}}$

Points to remember when constructing histograms are:

(*a*) the measurement (parameter) is *always* placed on the *x* axis ($\rightarrow$);
(*b*) the number of individuals is *always* placed on the *y* axis ($\uparrow$);
(*c*) axes should *fill* available graph paper and be *clearly labelled*;
(*d*) blocks (not lines) are constructed;
(*e*) a heading should be given, including parameter, units, numbers and species.

DEVELOPMENT AND REPRODUCTION
4.7 Selection and evolution

4

Natural selection

Where variety exists in a species, there is a range of characteristics which may *benefit* or *handicap* an individual by making it better or worse adapted to its environment. The environment or 'nature' selects which individuals survive and therefore the future of a species. This is called *natural selection* and is now thought to be the basis of organic evolution. A poorly adapted individual is less likely to live to reproductive age and pass on its characteristics in genes to the next generation. Its 'poor' genes are lost from the species' *gene pool*.

Example 1 Early flowering
Bluebells and primroses grow at ground level in early spring. As soon as trees come into leaf there is not enough light for plants to grow beneath them. Any variations that have not flowered by then will not reproduce sexually.

Example 2 Thrush's anvil
A thrush's anvil is a stone habitually used by a thrush to crack open the shells of the snails it eats. The empty broken shells that build up tell us which types of snail have been eaten. The colour and number of stripes on the shell are important. Those best camouflaged will not be seen by thrushes and eaten before they can reproduce. In grass on sand dunes, striped yellow-shelled snails survive predators better. On gravel, plain brown-shelled snails survive better.

Example 3 Industrial melamism of the Peppered Moth (*Biston betularia*)
The pale peppered moth (*Biston betularia*) is well camouflaged on pale, lichen-covered (unpolluted) tree trunks where it rests by day, largely unseen by the birds that prey on it. As a result of a single gene mutation that proved dominant, a dark form arose. Black moths survive better on bare, sooty (polluted) tree trunks in industrial areas, because here they are better camouflaged against predators, as shown below. Original research on this was carried out by *Kettlewell*.

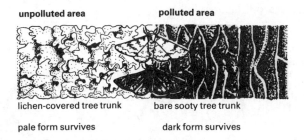

unpolluted area   polluted area

lichen-covered tree trunk   bare sooty tree trunk

pale form survives   dark form survives

Artificial selection

Plant and animal breeding This has been carried out by man for centuries. Plant and animal breeders with a population of a species at their disposal, have selected those with 'good' qualities and allowed them to breed, while those with 'poor' qualities have been killed or otherwise stopped from reproducing. Organisms produced in this way are:

(a) *Farm animals,* e.g. cows yielding more milk, sheep with more wool;
(b) *Pets,* e.g. pedigree varieties of cats, dogs, budgerigars;
(c) *Crops,* disease-resistant and high-yield (e.g. wheat, rice);
(d) *Garden plants,* e.g. varieties of roses, dahlias, lettuces.

Genetic engineering Recent advances have made it possible to transfer genes coding for useful properties from one bacterium to another. Such organisms are in use in the manufacture of drugs, vitamins, hormones and plastics. The process is known as genetic engineering and it is likely to play an enormous role in our industrial future.

Genetic counselling As more is discovered about genetic diseases, and as genetic engineering develops, such *counselling* or *advice* will become more common. Already, humans with a family history of a genetic disease can find out how likely they are to pass it on to their children. If the risk is high, they may decide to foster or adopt children, or consider using another person's egg or sperm to produce a 'test-tube baby'.

Evidence of evolution

Remembering the letters **C, D, E, F, G,** will remind you where evidence for evolution can be found:

Comparative anatomy;
Direct observation of species change;
Embryos;
Fossil records;
Geographical distribution.

Comparative anatomy

Homologues – similar structures possessed by organisms, having different functions (e.g. pentadactyl limb).

Vestiges – structures possessed by organisms, which have no apparent function, for example:
(a) Ostrich has full wing pattern but does not fly.
(b) Whale has limb bones in 'paddles'.
(c) Adder has pelvis and hind limb bones.
(d) Human being has appendix and third eyelid.

Direct observation of species change

Industrial melanism   The pale peppered moth is replaced by the melanic dark form in industrial (polluted) areas.

Selective breeding   Man has selected features of other organisms that are of advantage to him and has, by breeding programmes, developed new types of plant and animal (e.g. wheat, dogs).

Resistance   Frequent mutations in bacteria and other small organisms have been seen. This explains new resistant types of organism arising (e.g. rats resistant to warfarin, bacteria resistant to penicillin).

Embryos

Early embryos of *all* vertebrates are very similar. Up to a point, an individual's embryological development repeats its evolutionary development (e.g. the human embryo has gill pouches in its neck region).

Fossil records

Palaeontology is the study of *fossils*, left behind by dead plants and animals as evidence of their existence. Fossils may take several forms:

(a) *Whole preservation*, e.g. mammoth in ice, insects in amber;

(b) *Petrification*, e.g. trees – hard tissue replaced by 'stone';
(c) *Impressions*, e.g. *Archaeopteryx*, footprints – dents in silt or clay, which harden;
(d) *Casts*, e.g. ammonites – the shape of the organism is left in hardening clay after its death and decay. A second substance fills this mould to form the cast.

casts

The oldest fossils are dated by the rock in which they occur. Particular types of rock were laid down during definite periods, e.g. coal was formed during the carboniferous period 200–220 million years ago. More recent fossils can be dated by the *carbon-14 method* or the *potassium-argon method*. Both rely on the fact that organisms contain a constant proportion of *radioactive* substances which 'decay' after death at a constant rate.

Geographical distribution
Places of almost identical climates have very different plants and animals, e.g. penguins in Antarctica and polar bears in Arctica. Australia, which has most of the world's marsupials today, was separated from Asia before placental mammals originated.

Theories of evolution

Darwin (1809–1882) put forward the 'Theory of Natural Selection', summarised below:

(a) The number of potential offspring for exceeds the number which actually develop into adults.
(b) Since only a small proportion of offspring can survive, there is a *struggle for existence*.
(c) Variation exists between offspring.
(d) Offspring with characteristics giving an advantage in the struggle for existence stand a better chance of survival, i.e. *survival of the fittest*.

Lamarck (1744–1829) put forward the 'Law of Use and Disuse'. He held that the characteristics of an individual could change in its lifetime and that such changes could be passed on to future generations, e.g. that a giraffe with a neck stretched through reaching for leaves would have long-necked offspring. Likewise, he thought that an organ not used would become smaller and eventually vestigial.

DEVELOPMENT AND REPRODUCTION
Sample questions

4

1. The best parameter for measurement of growth in plants is
   A increase in dry weight
   B increase in volume
   C increase in number of nuclei
   D increase in cell wall material

2. Secondary growth in a flowering plant results from the production of
   A cork cells
   B new cambium
   C new phloem cells
   D new pith cells

3. Plant and animal growth are both controlled by
   A the brain B mineral salts C enzymes D hormones

4. A hydrotropism is a plant growth movement made in response to
   A light B gravity C water D touch

5. A tropism in a root results from extra growth in
   A meristems B vacuolating cells C the root cap D the cambium

6. Meiosis takes place in a flower within the
   A ovum B anther C filament D pollen

7. Mitosis takes place in a mammal within the
   A skin B ovary C sperm-producing cells D teeth

8. The best description of a seed is
   A an ovule
   B embryo and food in testa
   C a plumule and radicle
   D a flowering plant spore

9. Successful fertilisation of an egg by a sperm and its attachment to the uterus wall of a mammal is called
   A implantation B a zygote C conception D contraception

10. A gene is a
    A chromosome
    B characteristic
    C factor causing brown eyes
    D part of a chromosome

11. Identical twins are the result of the fusion of nuclei from
    A one ovum and one sperm
    B two ova and two sperms
    C two ova and one sperm
    D one ovum and two sperms

12. Which of the following could be given as evidence of evolution?
    A comparative anatomy B fossils C embryos D all of these

13. Fertilisation of a mammalian ovum usually occurs in the
    A ovary B oviduct C cervix D vagina

14. The hormone produced by male humans after puberty is
    A oestrogen B progesterone C testosterone D prolactin

15. Which of the following vertebrate groups show least parental care?
    A fishes B reptiles C birds D mammals

16 The three graphs below show the average daily hormone levels in the blood of 5 women during a menstrual cycle.

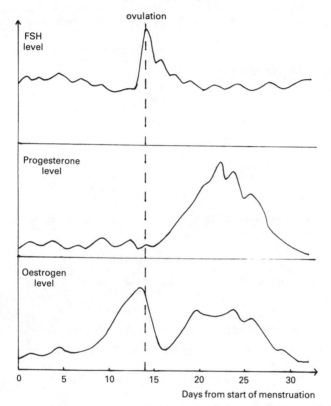

(a) What changes in hormone levels seem to trigger ovulation? (2)
(b) How true is it to say that as oestrogen level falls, the FSH level rises? (1)
(c) Comment on the validity of these data. (2)
(d) One type of contraceptive pill is taken for 21 days out of 28. The pill keeps the progesterone and oestrogen levels high. These levels fall during the 7 days when the pill is not taken. Using information from the graphs and your own knowledge, explain how the pill prevents fertilisation taking place. (5)

17 Some people are able to roll their tongues. These people are described as 'rollers'. Those who cannot roll their tongues are described as 'non-rollers'. The letter **R** represents the allele which gives rise to rollers and the letter **r** represents the allele which gives rise to non-rollers.

If Mrs Smith's genotype is **Rr** and Mr Smith's is **RR** explain why

(a) they are both rollers (2)
(b) all their children are rollers. *(Use the layout below.)* (2)

(4)
(1)

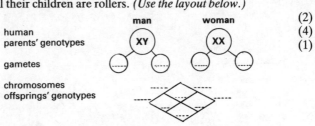

18 Write the symbols for the X and Y chromosomes in the circles below. (2)

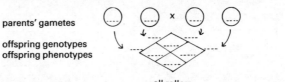

all rollers

(b) Fill in the Punnett square to show offsprings' genotypes (4)

19 (a) How does the placenta support the life and development of the embryo? (10)
(b) Describe the hazards to which the embryo in the uterus may be subjected. (5)
(c) Describe the process of childbirth. (5)

ANSWERS

Chapter One

1B 2A 3D 4C 5C 6C 7D 8A 9B 10B 11D 12C 13A 14B 15B	(15)
16 (a) A Insect, B Arachnid, C Insect, D Crustacean, E Myriapod	(5)
(b) Exoskeleton, jointed legs, have no backbone (any 2)	(2)
(c) It has been drawn ten times larger than life size.	(1)
(d) Pairs of wings on thorax, compound eyes	(2)
17 (a) (i) Having only one cell. (ii) Lacking cell cross walls (iii) Association/relationship benefiting both partners.	(3)
(b) (i) Protists. (ii) Monerans.	(2)
(c) No green chlorophyll, do not photosynthesise, no cellulose cell walls, nitrogen in cell walls, aseptate. (any 3)	(3)
(d) (i) Gullet, flagellum, eye spot, myoneme, contractile vacuole. (ii) Starch grains, chloroplasts.	(7)

18
Plants are useful –	*Animals* are useful –
in decay	as food source
as food sources	in providing: fibres, leather
in food industry e.g. brewing, baking, suspension	skin, cosmetics, drugs
	as pets
as pollution indicators	in working for man, e.g. ploughing, guarding property.
in providing: fibre, rubber, gum, drugs, wood or products	as biological control agents
	in improving soil fertility (10)
in maintaining atmospheric O_2 (10)	

(*One* mark for each *named* example of the above, up to a maximum of 10 for plants and 10 for animals)

Chapter Two

1B 2B 3A 4B 5C 6D 7A 8B 9C 10C 11D 12A 13B 14C 15A	(15)
16 (a) A piece of land –	(1)
put aside for man's recreation and enjoyment.	(1)
(b) (i) Provides somewhere to enjoy leisure time.	(1)
(ii) Provides a habitat for plants and animals.	(1)
(iii) Costs less/is easier.	(1)
(iv) Unsightly/unsafe	(1)

(c) (i) Giant panda, white rhino, orchids, or any other. (1)
(ii) Bamboo destroyed, animal overhunted, or habitat destroyed. (1)
(iii) Plant bamboo, establish nature reserves, makes laws, breed in zoos. (*Any 2*) (2)

17 (a) (i) 12% (ii) 1% (2)
(b) Percentage difference between smokers and non-smokers is greater in each town than – (1)
the difference between non-smokers in the two towns. (1)
(c) More sufferers are found in **A**, the polluted town. (1)
(d) Increases chance of lung cancer, paralyses cilia in breathing passages, blackens cilia, increases mucus secretion, causes emphysema. (*Any 3*) (3)

18 (a) Farming and agriculture have altered the appearance of the environment, affected communities living there and interfered with natural cycles, e.g. by addition of NPK fertilisers, combustion.
Building, industry and associated mining and pollution have removed and added materials circulating in the environments.
Hunting, fishing and deforestation have depleted natural resources.
Afforestation and nature reserves have restored them in places. (10)
(b) Population growth follows sigmoid curve (or draw graph) (2)
(c) Use of contraception, e.g. pill, sheath, IUD, spermicides, withdrawal, safe period and male and female sterilisation. (Do not include much detail) (8)

Chapter Three

1C 2B 3D 4A 5A 6D 7B 8D 9C 10C 11B 12A 13B 14D 15A 16C 17C 18B 19A 20D (20)

21 (a) Acid, neutral, alkali (3)
(b) Pepsin, salivary amylase, trypsin (3)

22 (a)

Animal	*Plant*
0.01mm diameter	0.1mm diameter
nucleus near middle	nucleus near side
rounded shape	angular shape
no chloroplasts	chloroplasts
no wall present	wall present
no permanent vacuole	central vacuole

(*Any 4*) (4)

(b) (i) Cell A is round, cell C is very long and thin; cell A has no extensions, cell C has dendrons. (2)
(ii) Length helps fast transmission of impulses; Dendrons link cell C with other nerve cells. (2)
(c) (i) Carries out photosynthesis. (1)
(ii) Palisade mesophyll in leaf. (2)
(d) (i) Carries out respiration. (2)
(ii) Muscle of animals. (2)

23 (a) Photosynthesis, using sunlight (2)
(b) Green pond animals (1)
(c) Other animals. (1)
(d) CO_2 level would drop (animals not respiring). (1)
Amount of plant material would increase initially (animals not eating them). (1)
and then decrease (lack of CO_2 for photosynthesis). (1)
(e) Yes. (1)
If enough saprophytes were present to recycle raw materials and a balance existed between the processes of respiration (1)
and photosynthesis (1)
it is possible for life to continue indefinitely. (3)

24 (a) 700 cm^3 (1)
(b) 3,750 cm^3 (5,000 cm^3 – 1,250 cm^3) (1)
(c) The amount left in the lungs after full expiration (1)
(d) The amount extra that can be inspired above normal inspiration volume. (1)
(d) 2,200 cm^3 (1)

25 (a) (**A**) Sunlight energy. (**B**) Carbon dioxide. (**C**) Water. (3)
(b) (**D**) Oxygen. (1)
(c) It has areas that contain chlorophyll and areas that do not.
(d) (i) It is made of very narrow tubes, no cross walls (*or any other suitable answer*). (1)
(ii) Water enters the guard cells when their osmotic pressure rises. This bends the cells. (1)
(e) To make chlorophyll. (1)

26 (a) Large surface area, thin to aid diffusion, moist to aid diffusion, well-ventilated, close to transport systems. (5)
(b) (i) Diaphragm down; ribs up and out; volume in chest increases; air drawn into alveoli; oxygen dissolves in water film; concentration gradient exists; oxygen diffuses into blood; carbon dioxide diffuses out. (5)

(ii) Diffusion occurs across a moist skin; when active air gulped into sac-like lung; no diaphragm. (5)
(iii) Has lungs and air sacs and passages; air is inhaled through nostrils; air passes twice through system to obtain maximum oxygen. (5)

Chapter Four

1A 2B 3D 4C 5B 6B 7A 8B 9C 10D 11A 12D 13B 14C 15A (15)

16 (a) Increases in oestrogen and FSH (1)
 (b) Not clearly true.
 (c) Not reliable – (1)
 as sample very small (5 women) (1)
 (d) Progesterone and oestrogen in pill mimic conditions between days 14 and 28 (1)
 They maintain uterus lining (1)
 as in pregnancy (1)
 and prevent ovulation, (1)
 which requires high oestrogen and low progesterone levels. (2)

17 (a) Both possess the dominant gene allele R, which shows in the phenotype. (2)
 (b)

$$\begin{array}{c} r \quad R \\ R \diagup Rr \diagdown R \\ RR \diagdown Rr \\ RR \end{array}$$

(4)

therefore all rollers (1)

18 (a)

(XY)   (XX)
(X) (Y) (X) (X)

(2)

 (b)

$$\begin{array}{c} Y \quad X \\ X \diagup XY \diagdown X \\ XX \diagdown XY \\ XX \end{array}$$

(4)

19 (a) Placenta comprises blood vessels from mother and embryo; supports embryo via umbilical cord; transports food, e.g. dissolved glucose, amino acids, vitamins, minerals and antibodies from mother to baby; removes waste urea and carbon dioxide from embryo, preventing poisoning; filters most harmful substances out of mother's blood before they reach embryo. (10)
 (b) Hazards include substances like that can cross the

placenta, (e.g. some drugs like nicotine, alcohol and thalidomide) and certain pathogens, e.g. rubella virus; mother's high blood pressure; any interference, e.g. amniocentesis, where needle is introduced into amnion; X-rays and other radiation. (5)

(c) 3 stages (i) Contractions of uterus muscles become stronger and more frequent, widen cervix, waters break and amnion leaks out. (ii) Head passes down vagina and baby is born; umbilical cord is cut. (iii) Placenta is delivered (afterbirth). (5)